Pressure and Temperature
Well Testing

Pressure and Temperature Well Testing

Izzy M. Kutasov

Department of Geosciences
Tel Aviv University, Tel Aviv, Israel

and

Lev V. Eppelbaum

Department of Geosciences
Tel Aviv University, Tel Aviv, Israel

CRC Press
Taylor & Francis Group
Boca Raton London New York

CRC Press is an imprint of the
Taylor & Francis Group, an **informa** business

A SCIENCE PUBLISHERS BOOK

CRC Press
Taylor & Francis Group
6000 Broken Sound Parkway NW, Suite 300
Boca Raton, FL 33487-2742

First issued in paperback 2019

© 2016 by Taylor & Francis Group, LLC
CRC Press is an imprint of Taylor & Francis Group, an Informa business

No claim to original U.S. Government works

ISBN-13: 978-1-4987-3361-8 (hbk)
ISBN-13: 978-0-367-37732-8 (pbk)

Visit the Taylor & Francis Web site at
http://www.taylorandfrancis.com

and the CRC Press Web site at
http://www.crcpress.com

Preface

The monograph is composed of two parts: Pressure and Flow Well Testing (Part 1) and Temperature Well Testing (Part 2). The logic for this composition is that: due to the similarity in Darcy's and Fourier's laws, the same differential diffusivity equation describes the transient flow of incompressible fluid in porous medium and heat conduction in solids. Therefore it is reasonable to assume that the techniques and data processing procedures of pressure well tests can be similarly applied to temperature well tests. Generally the mathematical model of pressure well tests is based on presenting the borehole as an infinite long linear source with a constant fluid flow rate in an infinite-acting homogeneous reservoir.

This model cannot be applied to stimulated wells (where the dimensionless production and shut-in time based on the apparent well radius can be very small).

In our work, the borehole is considered to be an infinite long cylindrical source with a constant or variable fluid flow rate. New semi analytical solutions are derived and utilized. In Part 1 presents:

1. New methods of determination formation permeability and skin factors from tests conducted in simulated wells
2. New procedures of processing data and designing interference well tests
3. New methods of determination of formation permeability and skin factor from afterflow pressure and sandface flow rate data
4. New methods of processing constant bottom-hole a step pressure tests

In Part 2 we discuss:

1. Several methods for estimation of the formation temperature and geothermal gradients from temperature surveys and logs
2. Determination *in-situ* the formation thermal conductivity and contact thermal resistance of boreholes
3. Cementing of casing in hydrocarbon wells: the optimal time lapse to conduct a temperature log
4. Temperature regime of boreholes: Cementing of production liners
5. Recovery of the thermal equilibrium in deep and super deep wells: Utilization of measurements while drilling data

The driving force at pressure well tests is the differential pressure—bottom-hole mud pressure minus the pressure of the formation fluid. For high temperature, high pressure (HTHP) wells, the hydrostatic pressure may significantly effect the differential pressure.

In the Appendix two equations are presented which can be used for estimation of the hydrostatic pressure. The early proposed empirical equation of state (the density-pressure-temperature dependence) for drilling fluids, brines and water was utilized in derivation of these two equations.

The monograph is based on many years of the authors' investigation, with the greatest emphasis being the last years publications: Kutasov (2003, 2007, 2013, 2015), Kutasov and Eppelbaum (2003, 2005, 2007, 2008, 2009, 2010, 2011, 2012a, 2012b, 2013a, 2013b, 2013c, 2014, 2015), Kutasov and Kagan (2000, 2001, 2003a, 2003b), Kutasov et al. (2008), Eppelbaum and Kutasov (2006a, 2006b, 2011, 2013), Eppelbaum et al. (1996, 2014).

The monograph is intended for students, engineers and researchers in the field of hydrocarbon geophysics and geology, groundwater searching and exploitation, and subsurface environment examination. It will be also useful for specialists studying pressure and temperature in parametric deep and superdeep wells.

Contents

Short Biographies of the Authors

 **Izzy M. Kutasov** holds a Ph.D. in Physics from O. Schmidt Earth Physics Institute in Moscow. He was a Senior Lecturer at the School of Petroleum Engineering, University of New South Wales, Sydney, and a graduate faculty member at the Department. of Petroleum Engineering and Geosciences, Louisiana Tech University. He worked for Shell Development Co., Houston, as a Senior Research Physicist. Now Dr. I. M. Kutasov is a consultant with BYG Consulting Co, Boston, USA. His research interests include the temperature regime of deep wells, transient pressure/flow analysis, well drilling in permafrost areas as well as deriving the climate of the past from subsurface temperature measurements. He is author of more than 250 publications including four books and more than 100 articles. His developments in the thermal data analysis in oil & gas wells received a wide recognition in the world scientific and engineering community. Dr. Kutasov has presented his scientific achievements at more than 60 International Conferences.

 **Lev V. Eppelbaum** received an MSc from the Azerbaijan Oil Academy in 1982, and a Ph.D. from the Geophysical Institute of Georgia in 1989. From 1982–1990 he worked as geophysicist, Researcher and Senior Researcher at the Institute of Geophysics in Baku (Azerbaijan). In 1991–1993 Eppelbaum completed his postdoctoral studies in the Dept. of Geophysics and Planetary Sciences at Tel Aviv University; at present he occupies a position of Associate Professor in the same Department. He is the author of more than 330 publications including 7 books, 125 articles and 55 proceedings. His scientific interests cover potential and quasi-potential geophysical field analysis, integrated interpretation of geophysical fields, tectonics and geodynamics. Eppelbaum's research interests in geothermics include examination of temperature regime of boreholes, thermal interactions at a depth, near-surface temperature measurements, study of climate of the past, and nonlinear analysis.

PART I

Pressure and Flow Well Testing

1

Pressure and Temperature: Drawdown Well Testing: Similarities and Differences

It is widely recognized now that borehole geophysical measurements are no less important than ground geophysical observations. Borehole geophysical logging is used for searching various economic minerals, fresh and hot water, tectonic-structural investigations and environmental and ecological analysis (Gretener 1981; Jorden and Campbell 1984; Bourdarot 1998). Temperature and pressure investigations fall within the domain of the most exploitable physical parameters (Serra 1984; Tittman 1986). The knowledge of thermal properties of formations (Kappelmeyer and Haenel 1974; Somerton 1992; Vosteen and Schellschmidt 2003) and initial formation temperature are needed to evaluate the energy capacity of geothermal reservoirs. The forecasting of fluid flow rate of production and injection geothermal wells requires an estimation of mobility (formation permeability and fluid viscosity ratio), porosity, total formation compressibility, skin factor, and initial reservoir pressure (Earlougher 1977; Elder 1981). In petroleum and geothermal reservoir engineering pressure and flow well tests are routinely conducted to determine these parameters (Earlougher 1977; Lee 1982; Prats 1982; Edwards et al. 1982; Sabet 1991; Horne 1995; Kutasov 1999). Due to the similarity in Darcy's and Fourier's laws the same differential diffusivity equation describes the transient flow of incompressible fluid in porous medium and heat conduction in solids. As a result, a correspondence exists between the following parameters: volumetric flow rate, pressure gradient, mobility, hydraulic diffusivity coefficient and heat flow rate, temperature gradient, thermal conductivity and thermal diffusivity. Thus, it is reasonable to assume that similar to the techniques and data processing procedures of pressure well tests can be applied to temperature well tests (Muskat 1946; Carslaw and Jaeger 1959). However, as it will be shown below, this approach can be used only in some cases (large dimensionless times). Generally the mathematical model of pressure well tests is based on presentation of the borehole as an infinite long linear source with a constant fluid flow rate in an infinite-acting homogeneous reservoir. For this case the well-known solution of the differential diffusivity equation is expressed through the exponential integral (Carslaw and Jaeger 1959). At temperature well testing the borehole (or the cylindrical heater) cannot be considered as an infinite long linear source of heat. This is due to low values of thermal diffusivity of formations (in comparison with the hydraulic diffusivity) and corresponding low values of dimensionless time. As was shown in (Kutasov 2003) the convergence of solutions of the diffusivity equation for cylindrical and linear sources occurs at dimensionless time of about 1000.

Earlier we suggested a semi-theoretical equation to approximate the dimensionless heat flow rate from an infinite cylindrical source with a constant bore-face temperature (Kutasov 1987a). This equation was used to process data of pressure and temperature well tests and to develop a technique for determining the formation permeability, skin factor, thermal conductivity and thermal resistance of the borehole (Kutasov 1998; Kutasov and Kagan 2003a, 2003b; Kutasov and Eppelbaum 2005; Eppelbaum et al. 2014).

The objective of this chapter is twofold: to discuss the similarities and differences in techniques of pressure and temperature well testing at a constant heat (fluid flow) rate. In our paper (Eppelbaum and Kutasov 2006b) we suggested working formulas for a drawdown temperature well test, a simulated example was also presented to demonstrate the data processing procedure for determination of formation thermal conductivity, initial temperature, skin factor, and contact thermal resistance.

1.1 Mathematical Models

1.1.1 Well as a cylindrical source

Let's assume that a well is producing at a constant flow rate from an infinite-acting reservoir and: reservoir is a homogeneous and isotropic porous medium of uniform thickness, porosity and permeability are constant, fluid of small and constant compressibility, constant fluid viscosity, small pressure gradients, negligible gravity forces and laminar flow in the reservoir and the Darcy law can be applied (Muskat 1946; Matthews and Russell 1967). For a well as a cylindrical source it is necessary to obtain the solution of the diffusivity equation under following boundary and initial conditions:

$$p(t=0,r) = p_i, \qquad r_w \leq r < \infty, \qquad t > 0, \tag{1-1}$$

$$\left(r\frac{\partial p}{\partial r}\right)_{r_w} = \frac{q^*\mu}{2\pi k}, \qquad q^* = \frac{q_h}{h}, \qquad t > 0, \tag{1-2}$$

$$p(t, r \to \infty) \to p_i, \qquad t > 0, \tag{1-3}$$

where q^* is fluid flow rate per unit of length, q_h is fluid flow rate, h is the net formation thickness, k is the permeability, r is the radial coordinate, h is the formation thickness, and p_i is the initial pressure.

It should be noted (Eq. (1-2)) that, due to convention, in reservoir engineering the produced flow rate (q_h) has a positive sign and negative sign is applied to injection wells. It is well known, that in this case, the diffusivity equation has a solution in complex integral form (Van Everdingen and Hurst 1949; Carslaw and Jaeger 1959; Bejan 2004). For the bottom-hole pressure (in terms of pressure and fluid flow rate) we obtained the following semi-analytical equation (Kutasov 2003).

$$p_{wf}(t) = p(t, r_w) = p_i - \frac{q^*\mu}{2\pi k} \ln\left[1 + \left(c - \frac{1}{a + \sqrt{t_D^*}}\right)\sqrt{t_D^*}\right],$$
(1-4)

$$t_D^* = \frac{kt}{\phi c_t \mu r_w^2}, \qquad a = 2.7010505, \qquad c = 1.4986055.$$
(1-5)

t_D^* is the dimensionless time, t is the time, μ is the viscosity, ϕ is the porosity, c_t is the total compressibility, and p_{wf} is the flowing bottom-hole pressure.

Similarly, for the borehole as a cylindrical source of heat ($r_h = r_w$) the wall temperature is

$$T_w = T(t, r_w) = T_i + \frac{q}{2\pi\lambda} \ln\left[1 + \left(c - \frac{1}{a + \sqrt{t_D}}\right)\sqrt{t_D}\right],$$
(1-6)

$$t_D = \frac{\chi t}{r_w^2} = \frac{\lambda t}{c_p \rho r_w^2}.$$
(1-7)

Here T_i is the initial (undisturbed) formation temperature, T_w is the wall temperature, q is the heat flow rate per unit of length, λ is the thermal conductivity of formations, χ is the thermal diffusivity of formations, ρ is the density of formations, and c_p is the specific heat at constant pressure.

Let's introduce the dimensionless wall pressure and dimensionless wall temperature

$$p_D(t_D^*) = \frac{2\pi k h(p_i - p_w)}{q_h \mu},$$
(1-8)

$$T_D(t_D) = \frac{2\pi\lambda(T_w - T_i)}{q}.$$
(1-9)

Then

$$p_D(t_D^*) = \ln\left[1 + \left(c - \frac{1}{a + \sqrt{t_D^*}}\right)\sqrt{t_D^*}\right],$$
(1-10)

$$T_D(t_D) = \ln\left[1 + \left(c - \frac{1}{a + \sqrt{t_D}}\right)\sqrt{t_D}\right].$$
(1-11)

Values of T_D calculated from Eq. (1-11) and results of a numerical solution ("Exact" solution) by Chatas (Lee 1982) were compared (Kutasov 2003). The agreement between values of T_D calculated by these two methods was very good. For this reason the principle of superposition can be used without any limitations.

1.1.2 Well as a linear source

It is clear from physical considerations that, for large values of dimensionless time, the solutions for cylindrical and linear sources should converge. To develop the solution for a linear source, the boundary condition expressed by Eq. (1-2) should be replaced by the condition:

$$\lim_{r \to 0}\left(r\frac{\partial p}{\partial r}\right) = \frac{q^*\mu}{2\pi k}, \qquad q^* = \frac{q_h}{h}, \qquad t > 0 \tag{1-12}$$

and the well-known solution for the pressure (at $r = r_w$) for an infinitely long linear source with a constant flow rate in an infinite-acting reservoir is (Carslaw and Jaeger 1959)

$$p_i - p_{wf} = \frac{q^*\mu}{2\pi k}\left[-\frac{1}{2}Ei\left(-\frac{1}{4t_D^*}\right)\right], \qquad p_i > p_{wf}, \tag{1-13}$$

where Ei is the exponential integral.

$$p_D(t_D^*) = -\frac{1}{2}Ei\left(-\frac{1}{4t_D^*}\right). \tag{1-14}$$

We compared function p_D (Eq. (1-14)) and the "Exact" solution of Chatas (Lee 1982) and found (Kutasov 2003) that at about $t_D^* > 500$ the Ei-function approximates the bottom-hole pressure with good accuracy (Figure 1-1).

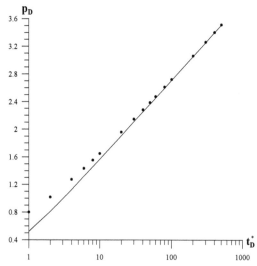

Figure 1-1. Dimensionless pressure versus dimensionless time. Solid line – Eq. (1-14), points – "Exact" solution of Chatas (Lee 1982).

The corresponding equations in terms of temperature and heat flow rate are

$$T_i - T_w = -\frac{q}{2\pi\lambda}T_D(t_D), \qquad T_w > T_i, \tag{1-15}$$

$$T_D(t_D) = -\frac{1}{2}Ei\left(-\frac{1}{4t_D}\right). \tag{1-16}$$

1.1.3 The corresponding parameters

We mentioned in the Introduction that the same differential diffusivity equation describes the transient flow of incompressible fluid in porous medium and heat conduction in solids. As a result, a correspondence exists between the parameters presented in Table 1-1.

Table 1-1. Correspondence between the flow of an incompressible fluid through a porous medium and heat conduction in solids. Borehole—formation system.

Steady-state flow through porous medium	Heat conduction
Pressure: p	Temperature: T
Pressure gradient: ∇p	Temperature gradient: ∇T
Initial reservoir pressure: p_i	Undisturbed formation temperature: T_i
Fluid flow rate: $q^* = -k/\mu\nabla p$, (Darcy's law)	Heat flow rate: $q = -\lambda\nabla T$, (Fourier's law)
Permeability viscosity ratio: k/μ	Thermal conductivity: λ
Porosity and total compressibility product: ϕc_t	Density and specific heat product: ρc_p
Hydraulic diffusivity: $\eta = k/\mu\phi c_t$	Thermal diffusivity: $a = \lambda/\rho c_p$
Well radius: r_w	Heater radius: r_h
Radius of drainage: r_{dr}	Radius of thermal influence: r_{in}
Dimensionless parameters	
Radial distance: $r_D = r/r_w$	Radial distance: $r_D = r/r_h$
Radius of drainage: $R_{dr} = r_{dr}/r_w$	Radius of thermal influence: $R_{in} = r_{in}/r_h$
Time: $t_D^* = \eta t/r_w^2$	Time: $t_D = at/r_h^2$
Pressure: p_D	Temperature: T_D

It will be shown (field example) that for pressure well tests are typical high values of hydraulic diffusivity coefficient. This leads to very high values of dimensionless time, and the borehole can be considered as a linear source. We should also note that porosity and total compressibility product (ϕc_t) cannot be determined from a test in one well and only at multi-well testing (interference testing) this parameter can be obtained. Fortunately, the analogous parameter—density and specific heat product (ρc_p) vary within narrow limits and can be determined from cuttings (Kappelmeyer and Haenel 1974; Somerton 1992).

1.2 Drainage and Thermal Influence Radius

Presented solutions of the diffusivity differential equation are valid only for the transient (infinite-acting reservoir) period. This means that during the test the pressure (or temperature) field around the borehole is practically not affected by reservoir's boundaries or by others production (injection) wells. Therefore, in order to estimate the duration of the transient period, the drainage (thermal influence) radius should be determined with a sufficient accuracy. Below we present a general approach that allows one to determine the drainage (thermal influence) radius as a function of time for a well (cylindrical heat source) produced at a constant flow (heat) rate.

1.2.1 Drainage radius

It is desirable to have an approximate relationship between dimensionless cumulative production and dimensionless time (Kutasov and Hejri 1984; Johnson 1986). We will use the material balance condition to determine the well drainage radius. The corresponding general equation for the dimensionless drainage radius R_{dr} is (Kutasov 1999):

$$Q_D^* = \int_1^\infty p_{Dr}\left(r_D, t_D^*\right) r_D \, dr_D = \int_1^{R_{dr}} \left(1 - \frac{\ln r_D}{\ln R_{dr}}\right) r_D \, dr_D, \qquad R_{dr} > r, \tag{1-17}$$

$$p_{Dr}\left(r_D, t_D^*\right) = \frac{p(r,t) - p_i}{p_{wf} - p_i}, \qquad r_D = \frac{r}{r_w}, \qquad R_{dr} = \frac{r_{dr}}{r_w}, \tag{1-18}$$

where Q_D^* is the dimensionless cumulative fluid production per unit of length and r_{dr} is the radius of drainage.

Using tables of integrals (Gradstein and Ryzhik 1965) we obtain

$$Q_D^* = \frac{1}{4} \frac{R_{dr}^2 - 2\ln(R_{dr}) - 1}{\ln(R_{dr})}. \tag{1-19}$$

The cumulative fluid production per unit of length is

$$Q^* = q^* t = 2\pi \phi c_t r_w^2 \left(p_i - p_{wf}\right) Q_D \tag{1-20}$$

Combining Eqs. (1-4), (1-5), (1-19), and (1-20) we obtain an equation to calculate the value of R_{dr}

$$t_D^* = \frac{1}{4} \frac{R_{dr}^2 - 2.\ln(R_{dr}) - 1}{\ln(R_{dr})} \cdot \ln\left[1 + \left(c - \frac{1}{a + \sqrt{t_D^*}}\right)\sqrt{t_D^*}\right]. \tag{1-21}$$

1.2.2 Thermal influence radius

In theory a cylindrical source with a constant heat flow rate affects the temperature field of formations at very long radial distances. There is, however, a practical limit to the distance—the radius of thermal influence (r_{in}), where for a given period the temperature $T(r_{in},t)$ is "practically" equal to the geothermal temperature T_i. To avoid uncertainty, however, it is essential that the parameter r_{in} must not to be dependent on the temperature difference $T(r_{in},t) - T_i$. For this reason we used the thermal balance method to calculate the radius of thermal influence. Similarly, we obtain:

$$Q_D = \int_1^\infty T_{Dr}(r_D,t_D)r_D dr_D = \int_1^{R_{in}}\left(1 - \frac{\ln r_D}{\ln R_{in}}\right)r_D\, dr_D, \qquad R_{in} > r_D, \tag{1-22}$$

$$T_{Dr}(r_D,t_D) = \frac{T(r,t) - T_i}{T_w - T_i}, \qquad R_{in} = \frac{r_{in}}{r_w}, \tag{1-23}$$

where Q_D is the dimensionless cumulative heat production per unit of length.

Similarly,

$$Q_D = \frac{1}{4}\frac{R_{in}^2 - 2\ln(R_{in}) - 1}{\ln(R_{in})}. \tag{1-24}$$

The cumulative heat flow per unit of length is given by

$$Q = qt = 2\pi c_p r_w^2 (T_w - T_i)Q_D. \tag{1-25}$$

Combining Eqs. (1-6), (1-7), (1-24) and (1-25) we obtain an equation to determine the value of R_{in}

$$t_D = \frac{1}{4}\frac{R_{in}^2 - 2\ln(R_{in}) - 1}{\ln(R_{in})} \cdot \ln\left[1 + \left(c - \frac{1}{a + \sqrt{t_D}}\right)\sqrt{t_D}\right]. \tag{1-26}$$

1.3 Skin Factor and Borehole Storage

1.3.1 Fluid flow

As a consequence of drilling and completion operations most wells have reduced permeability near the borehole (skin zone). This results in pressure loss due to skin and corresponding reduction in well productivity. Well stimulation through acidicing or hydraulic fracturing may improve the well productivity even above the production levels corresponding to undamaged conditions. Quantitatively the skin is defined as a parameter (skin factor) which depends on the thickness and the effective permeability of the skin zone (Hawkins 1956):

$$s^* = \left(\frac{k}{k_s} - 1\right)\ln\frac{r_s}{r_w}, \tag{1-27}$$

where k_s is the permeability of the skin zone, and r_s is the radius of the skin zone.

It is more convenient to express the skin factor through the apparent (effective) well radius, r_{wa} (Earlougher 1977):

$$r_{wa} = r_w e^{-s^*}.$$

(1-28)

Introducing the pressure drop due to skin into Eq. (1-13), we obtain

$$p_i - p_{wf} = \frac{q_h \mu}{2\pi kh} \left[-\frac{1}{2} Ei \left(-\frac{1}{4t_D^*} \right) + s^* \right].$$

(1-29)

For large values of t_D^* the logarithmic approximation of the Ei function can be used, and

$$-\frac{1}{2} Ei \left(-\frac{1}{4t_D^*} \right) \approx \frac{1}{2} \left(\ln t_D^* + \ln 4 - 0.57722 \right) = \frac{1}{2} \left(\ln t_D^* + 0.80907 \right),$$

(1-30)

where 0.57722 is the Euler's constant.

The last equation for values of $x < 0.001$ approximates the $Ei(-x)$ function with an error less than 0.016% and with absolute accuracy of 0.001. It is assumed (Eq. (1-13)) that during the pressure drawdown test the flow rate from the reservoir into the borehole (the "sand face" flow rate, q_{sf}) is constant for $t > 0$. In most of the cases the flow rate is controlled and measured at the wellhead. When the borehole is completely full of a single-phase fluid, for some time the fluid stored in the borehole will provide a fraction of the flow rate. During this period the "sand face" flow rate will increase from $q_{sf}(t = 0) = 0$ to $q_{sf} = q_h$. It is commonly assumed this increase can be approximated by the exponential function (Kuĉuk and Ayestaran 1985a,b):

$$q_{sf} = q_h \left(1 - e^{-\alpha t} \right), \qquad \alpha > 0.$$

(1-31)

Horne (1995) gives a comprehensive description of the borehole storage phenomena for a pressure drawdown test: *When the well is first open to flow, the pressure in the borehole drops. This drop causes an expansion of the borehole fluid, and thus the first production is not fluid from the reservoir, but is fluid that had been stored in the borehole volume. As the fluid expands, the borehole is progressively emptied, until the borehole system can give up no more fluid, and it is the borehole itself which provides most of the flow during this period. This is borehole storage due to fluid expansion.*

1.4 Pressure Drawdown Well Testing

A drawdown test at a constant fluid flow rate is often used to determine formation permeability and the extent of the formation damage (or improvement) around production or injection wells. At present the oilfield (practical) units are usually used to measure and process field data. The dimensions of the parameters are: $[q]$ = standard barrel per day (STB/D), $[B]$ = RB/STB, $[k]$ = md (millidarcy), $[p]$ = psi (pound-force per squared inch), $[t]$ = hr, $[h]$ = ft, $[\mu]$ = cP, $[c_t]$ = 1/psi, $[r_w]$ = ft, $[\phi]$ = fraction, and $[\eta]$ = ft²/hr. STB is one barrel at standard conditions (p = 14.7 psi, T = 60°F) and RB is one barrel at reservoir pressure and temperature. Equations (1-29) and (1-30) may be combined and rearranged to familiar form (in oilfield units) of the pressure drawdown equation (Matthews and Russell 1967).

$$P_{wf} = P_i - \frac{162.6qB\mu}{kh}\left[\log t + \log\left(\frac{k}{\phi\mu c_t r_w^2}\right) - 3.2275 + 0.86859s *\right], \qquad (1\text{-}32)$$

$$P_{wf} = m\log t + P_{1hr}, \qquad (1\text{-}33)$$

$$m = -\frac{162.6qB\mu}{kh}, \qquad (1\text{-}34)$$

$$P_{1hr} = P_i + m\left[\log\left(\frac{k}{\phi\mu c_t r_w^2}\right) - 3.2275 + 0.86859s *\right]. \qquad (1\text{-}35)$$

As can be seen from Eq. (1-33) the slope of plot bottom-hole pressure versus logarithm of flowing (production) time allows to determine the formation permeability (Eq. (1-34)) and from the intercept (Eq. (1-35)) the skin factor can be calculated. In Eq. (1-35) the value of p_{1hr} must be taken from the semilog straight line. If pressure data measured at 1 hour do not fall on that line, the line must be extrapolated to 1 hour and the extrapolated value of p_{1hr} must be used (Earlougher 1977). The value of p_{1hr} (when log 1 = 0) is used in petroleum engineering to simplify Eq. (1-32). In practice many phenomena such as: insufficient stabilization time, the duration of wellbore storage, impact of reservoir boundaries, reservoir heterogeneities, partial well completion, transition from linear to radial flow and the accuracy of pressure recording gauges can affect the results of drawdown pressure well tests. The effect of these phenomena in the in determining of formation permeabilities and skin factors are discussed in literature (Matthews and Russell 1967; Earlougher 1977; Lee 1982; Sabet 1991; Horne 1995; Bourdarot 1998; Kutasov 1999; Eppelbaum et al. 2014). Below we will demonstrate that a similar technique (based on presentation of the borehole as an infinite linear source) cannot be used at temperature drawdown testing (small dimensionless time).

1.4.1 Field example (Earlougher 1977, Example 3.1)

A pressure drawdown test was conducted in an oil well. The reservoir and test data are presented in Table 1-2 and Figure 1-2. From Figure 1-2 follows that deviation from a straight line caused by the skin and borehole storage effects can be observed during the first 4 hours. The duration of the transient period is 12 hours and at $t > 16$ hours the pressure behavior is influenced by reservoir's boundaries, by interference from nearby wells, or other factors.

Table 1-2. Reservoir data.

$h = 130$ ft	$\phi = 0.20$
$r_w = 0.25$ ft	$\mu = 3.93$ cP
$q_h = 348$ STB/D	$m = -22$ psi/cycle (Figure 1-2)
$B = 1.14$ RB/STB	$P_{1hr} = 954$ psi (Figure 1-2)
$c_t = 8.74 \cdot 10^{-6}$ 1/psi	$p_i = 1{,}154$ psi

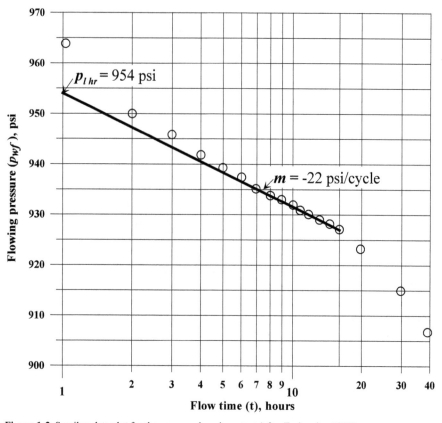

Figure 1-2. Semilog data plot for the pressure drawdown test (after Earlougher 1977).

From Eqs. (1-32)–(1-35) the following parameters were determined (Earlougher 1977):

$k = 89$ md and $s* = 4.6$. In the oilfield units the dimensionless time is

$$t_D^* = \frac{0.0002637\, kt}{\phi c_t \mu r_w^2}.$$ (1-36)

Let's calculate the dimensionless time at $t = 1$ hour

$$t_D^* \,(t = 1\,hr) = \frac{0.0002637 \cdot 89 \cdot 1}{0.20 \cdot 8.74 \cdot 10^{-6} \cdot 3.93 \cdot 0.25^2} = 5.47 \cdot 10^4.$$

As we noted earlier at value of dimensionless time of about 1000 the borehole can be considered as an infinite linear source. Now let's assume that a temperature drawdown test is conducted and: the formation is sandstone with the thermal diffusivity coefficient of $\chi = 0.00400$ m²/hr, radius of an electrical heater is also $r_w = 0.25$ ft, then

$$t_D(t = 1\,hr) = \frac{0.004 \cdot 1}{(0.25 \cdot 0.3048)^2} = 0.689.$$

Let's also evaluate the η/χ ratio

$$\begin{cases} \eta = \dfrac{0.0002637 \cdot 89}{0.20 \cdot 8.74 \cdot 10^{-6} \cdot 3.93} = 3.416 \cdot 10^3 \left(\dfrac{ft^2}{hr}\right) = 3.174 \cdot 10^2 \left(\dfrac{m^2}{hr}\right), \\[3mm] \dfrac{\eta}{\chi} = 3.174 \cdot 10^2 / 0.00400 = 7.93 \cdot 10^4 . \end{cases}$$

It is clear that at conducting a temperature drawdown test the cylindrical heater cannot be considered as an infinite long linear source of heat. We also calculated Eqs. (1-21) and (1-26) dimensionless values of drainage and thermal influence radiuses (Table 1-3).

After 10 hours the radius of the thermal influence is $r_{in} = 6.19 \cdot 0.25 = 1.548$ (ft) $= 0.472$ (m) and the drainage radius is $r_{dr} = 1512 \cdot 0.25 = 378$ (ft) $= 132$ (m).

Table 1-3. Drainage and thermal influence radius.

t, hours	R_{in} $t_D (t = 1\ hr) = 0.689$	R_{dr} $t_D (t = 1\ hr) = 5.47 \cdot 10^4$
1	2.57	480
2	3.26	678
3	3.79	830
4	4.23	958
5	4.63	1070
6	4.99	1172
7	5.32	1265
8	5.63	1352
9	5.92	1434
10	6.19	1512

Conclusion

The similarities and differences in techniques of pressure and temperature drawdown well testing are demonstrated. It is shown that at conducting a temperature drawdown test the cylindrical heater (probe) cannot be considered as an infinite long linear source of heat.

2

The Adjusted Circulation Time

In his classical paper H. Ramey (1962) had drawn the attention of petroleum engineers to the fact that the three solutions: for a cylinder losing heat at constant temperature, for a constant heat flowline source, and for a cylinder losing heat under the convection boundary, practically converge after some time. The simple solution for a constant fluid flow (or heat flow) line source is expressed through the exponential integral, and is widely used in reservoir engineering. To use this solution for any value of circulation time, we introduce below the adjusted circulation time. For the first time the concept of adjusted dimensionless production time was suggested by Ehlig-Economides and Ramey (1981). Let us assume that during time $0 \leq t \leq t_p$ the wellbore is producing at a constant bottom-hole pressure. In this case the radial pressure distribution and the time dependent flow rate are expressed by complex integrals. By using the material balance condition the adjusted dimensionless production time can be obtained from Ehlig-Economides and Ramey (1981):

$$t_D^* = \frac{N_D}{q_{LD}}.$$

The actual adjusted circulation time is

$$t_c^* = \frac{t_D^* r_w^2}{\eta},$$

where N_D is the dimensionless cumulative production and q_{LD} is the dimensionless flow rate at $t = t_p$ and η is the hydraulic diffusivity. Thus we replaced the fluid production at a constant bottom-hole pressure by a fluid production at a constant flow rate. And exponential integral solution for the pressure distribution and the flow rate can be utilized. Now let us assume that during the time period $0 < t \leq t_p$ the temperature of the drilling fluid at a given depth can be considered as a constant. In this case by using the heat energy balance condition the adjusted dimensionless heating time can be obtained from

$$t_D^* = \frac{Q_D}{q_L}.$$

The actual adjusted heating time is

$$t_c^* = \frac{t_D^* r_w^2}{\eta}.$$

Here Q_D is the dimensionless cumulative heat flow rate and q_L is the dimensionless heat flow rate at $t = t_p$, and a is the thermal diffusivity of formations. Thus we replaced the process of heating of formations at a constant temperature (at a given depth) by a process of heating at a constant heat flow rate.

The values of $N_D = Q_D$ and $q_{LD} = q_L$ (results of numerical of numerical solutions) are presented in the literature (Van Everdingen and Hurst 1949; Jacob and Lohman 1952; Edwardson 1962; Sengul 1983). Using these data we obtained (Kutasov 1987):

$$\left\{ \begin{array}{ll} G = \dfrac{t_c^*}{t} = \dfrac{t_D^*}{t_D} = 1 + \dfrac{1}{1+AF} & t_D \leq 10 \\[2ex] F = \left[\ln(1+t_D)\right]^n & n = 2/3 \quad A = 7/8 \end{array} \right\}, \tag{2-1}$$

$$G = \frac{t^*}{t} = \frac{t_D^*}{t_D} = \frac{\ln t_D - \exp\left(-0.236\sqrt{t_D}\right)}{\ln t_D - 1}, \qquad t_D > 10. \tag{2-2}$$

The correlation coefficient $G(t_D)$ varies in the narrow limits: $G(0) = 2$ and $G(\infty) = 1$.

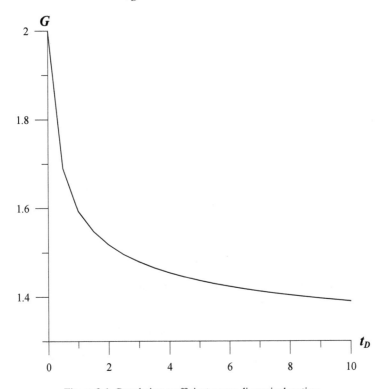

Figure 2-1. Correlation coefficient versus dimensionless time.

To simplify calculations we present Table 2-1.

Table 2-1. Function $G = G(t_D)$.

t_D	G	t_D	G	t_D	G
0.02	1.940	0.8	1.620	6.5	1.417
0.04	1.908	0.9	1.606	7.0	1.412
0.06	1.884	1.0	1.593	7.5	1.408
0.08	1.863	1.2	1.572	8.0	1.403
0.10	1.846	1.4	1.555	8.5	1.399
0.12	1.830	1.6	1.541	9.0	1.396
0.14	1.816	1.8	1.528	9.5	1.393
0.16	1.803	2.0	1.518	10.0	1.389
0.18	1.791	2.4	1.500	11.0	1.388
0.20	1.780	2.8	1.485	12.0	1.376
0.25	1.756	3.2	1.473	13.0	1.366
0.30	1.736	3.6	1.463	14.0	1.358
0.35	1.718	4.0	1.454	15.0	1.351
0.40	1.703	4.4	1.446	16.0	1.345
0.45	1.689	4.8	1.440	17.0	1.339
0.50	1.676	5.2	1.434	18.0	1.335
0.60	1.654	5.6	1.428	19.0	1.330
0.70	1.635	6.0	1.423	20.0	1.327

For large values of t_D the G function is given by

$$G = \frac{\ln t_D}{\ln t_D - 1}, \quad t_D > 1000. \tag{2-3}$$

Below we present two examples of calculation.

A. Production at a constant bottom-hole pressure: $t_p = 100$ hrs, well radius 0.1 m and hydraulic diffusivity = 300 m²/hr.

Step 1: Calculate the dimensionless production time $\dfrac{t_D(100 \cdot 300)}{0.1^2} = 3 \cdot 10^6$.

Step 2: From Eq. (2-2) we obtain $G = 1.072$ and $t^* = 1.072 \cdot 100 = 107.2$ (hrs).

B. Heating at a constant temperature of the drilling mud: $t_p = 100$ hrs, well radius 0.1 m and thermal diffusivity = 0.004 m²/hr.

Step 1: Calculate the dimensionless production time $\dfrac{t_D(100 \cdot 0.04)}{0.1^2} = 40$.

Step 2: From Eq. (2-2) we obtain $G = 1.288$ and $t^* = 1.288 \cdot 100 = 128.8$ (hrs).

3

Determination of the Formation Permeability and Skin Factor from a Variable Flow Rate Drawdown Test

3.1 Diffusivity and Boundary Conditions

A drawdown test at a constant flow rate is often used to determine formation permeability and the extent of the formation damage (or improvement) around production or injection wells. A control of the wellhead pressure is required to maintain a constant flow rate. In many drawdown tests the sandface flow rate is changing with time. Below we present a technique of processing data of VFRD tests. It is assumed that the flow of a single-phase fluid in an infinite reservoir is described with a diffusivity equation in cylindrical coordinates:

$$\frac{\partial^2 p}{\partial r^2} + \frac{1}{r}\frac{\partial p}{\partial r} = \frac{\phi \mu c_t}{k}\frac{\partial p}{\partial t}. \tag{3-1}$$

The use of this equation implies a number of conventional assumptions: isothermal flow of fluids of small and constant total compressibility c_t, constant porosity φ, permeability (k) and fluid viscosity μ, and the neglect of gravity forces. It is assumed that the initial reservoir pressure (p_i) is constant throughout the reservoir.

Let it be assumed also that during a variable flow rate drawdown (VFRD) test, the sandface flow rate is arbitrary and continuous function of time:

$$q(t) = q_{ref}\, q_D(t) \neq 0, \tag{3-2}$$

$$p(r,0) = p_i; \quad r > 0, \tag{3-3}$$

where q_{ref} is the reference bottom-hole flow rate. To determine the bottom-hole flowing pressure during the VFRD test it is necessary to obtain the solution of Eq. (3-1) with the following initial and boundary conditions:

$$q(t) = \frac{2\pi kh}{\mu}\left(r\frac{\partial p}{\partial r}\right)_{r_w}, \qquad p(\infty, t) = p_i,$$ (3-4)

where h is the reservoir thickness, and r_w is the wellbore radius.

3.2 Dimensionless Variables

The dimensionless time is defined by:

$$t_D = \frac{kt}{\phi\mu c_t r_w^2}.$$ (3-5)

The dimensionless time is very large and in practice the cylindrical source (borehole) is substituted by a line source and the exponential integral solution (*Ei*–function) can be used. The dimensionless wellbore pressure drop is defined by:

$$p_{wD} = \frac{p_i - p_{wf}(t)}{M}, \qquad M = \frac{q_{ref}\mu}{4\pi kh}.$$ (3-6)

Let us assume that dimensionless sandface flow rate $q_D(t)$ can be approximated by a polynomial of some degree n

$$q_D(t_D) = \sum_{i=0}^{n} a_i t^i,$$ (3-7)

$$q_D(t_D) = \sum_{i=0}^{n} b_i t^i,$$ (3-8)

$$b_i = \sum_{i=0}^{n} a_i \left(\frac{t}{t_D}\right)^i,$$ (3-9)

where $a_0, a_1, a_2, \ldots a_n$ are constant coefficients.

3.3 Duhamel Integral

It is well known that the superposition theorem (Duhamel integral) can be used to derive solutions for time-dependent boundary conditions (Carslaw and Jaeger 1959). In our case:

$$p_{wD}(t_D) = \int_0^{t_D} q_D(\tau) p_D'(t_D - \tau)\, d\tau,$$ (3-10)

where $p_D(t_D)$ is the dimensionless sandface pressure for the constant-rate case without the skin effect.

$$p_D'(t_D) = \frac{dp_D(t_D)}{dt_D},$$ (3-11)

$$p_D(t_D) = -Ei\left(-\frac{1}{4t_D}\right),$$ (3-12)

$$p_D'(t_D) = \frac{1}{t_D} \exp\left(-\frac{1}{4t_D}\right). \tag{3-13}$$

The skin effect can be included into Eq. (3-10) by adding the dimensionless pressure drop $2sq_D$ and considering Eq. (3-13) one obtains:

$$p_{wD}(t_D) = \int_0^{t_D} q_D(\tau) \exp\left[-\frac{1}{4(t_D-\tau)} t_D\right] \frac{d\tau}{t_D-\tau} + 2sq_D(t_D). \tag{3-14}$$

3.4 Working Formulas

Taking into account the conventional assumption that

$$t_D \gg 1, \quad \exp\left(-\frac{1}{t_D}\right) \approx 1 \tag{3-15}$$

and using the logarithmic approximation of the *Ei* function we obtained (Kutasov 1987):

$$\frac{p_i - p_{wf}}{q_D(t)} = M\left[\ln t - \frac{F(t,n)}{q_D(t)} + \ln\frac{4k}{\phi\mu c_t r_w^2} - 0.57722 + 2s\right], \tag{3-16}$$

where 0.57722 is the Euler's constant and

$$F(t,n) = \sum_{i=1}^{n} a_i t^i \sum_{j=1}^{i} \frac{1}{j}. \tag{3-17}$$

From a linear plot of $X = \ln t - F(t, n)/q_D(t)$ versus $Y = (p_i - p_{wf})/q_D(t)$ the values of the slope (M) and intercept (*Int*) can be obtained. The formation permeability is computed from Eq. (3-6), where 0.57722 is the Euler's constant. An example of data processing is presented in Table 3-2. The function $F = F(t, n)$ was calculated from Eq. (3-17) and skin factor is calculated from equation:

$$\frac{Int}{M} = \ln\frac{4k}{\phi\mu c_t r_w^2} - 0.57722 + 2s, \tag{3-18}$$

$$k = \frac{q_{ref} B\mu}{4\pi h M}. \tag{3-19}$$

3.4.1 Field example

Production rate during a 48-hour drawdown test (Earlougher 1977, Example 4.1) declined from 1,580 STB/D (10.47 m³/hr) to 983 STB/D (6.51 m³/hr). Reservoir data and well data are presented in Table 3-1 (to calculate the value of the skin factor we assumed the value of c_t and ϕ). For the first 7.5 hours rate decline can be approximated by a linear equation:

$$q(t) = q_{ref}(a_0 + a_1 t).$$

From linear regression analysis it was found that: q_{ref} = 1613.5 STB/D (10.69 m³/hr), a_0 = 1.00; a_1 = –0.0174 1/hr.

The values of X and Y were calculated (Table 3-2, Figure 3-1)

$$X = \ln t - \frac{a_1 t}{a_0 + a_1 t},$$

$$Y = \frac{p_i - p_{wf}}{a_0 + a_1 t}.$$

A linear regression computer program was used to determine the slope and intercept in following equation

$$Y = MX + Int.$$

Table 3-1. Input parameters.

h = 12.192 (m)	μ = 0.0006, Pa·s
ϕ = 0.20	p_i = 20.036, MPa
c_t = 1.305·10⁻⁸(1/Pa)	r_w = 0.1067, m
B = 1.27	

Table 3-2. Results of field data processing.

t, hrs	$p_i - p_{wf}$, MPa	X	Y, MPa
1.00	6.0881	8.2064	6.1959
1.50	6.4673	8.6210	6.6406
1.89	6.6534	8.8593	6.8797
3.00	6.9913	9.3424	7.3763
3.45	7.0602	9.4909	7.5111
3.98	7.1223	9.6444	7.6522
4.50	7.1637	9.7777	7.7722
5.50	7.2602	9.9993	8.0285
6.05	7.3291	10.1064	8.1914
6.55	7.3912	10.1968	8.3419
7.00	7.4188	10.2733	8.4477
7.50	7.4394	10.3537	8.5560

It was found that M = 1.0748·10⁶ Pa; Int = –2.6556·10⁶. The formation permeability is computed from Eq. (3-18):

$$k = \frac{(2.969 \cdot 10^{-3})(1.27)(6 \cdot 10^{-4})}{(4)(3.1415)(12.192)(1.0748 \cdot 10^{6})} = 1.374 \cdot 10^{-14} \, (m^2) = 13.92 \, (md).$$

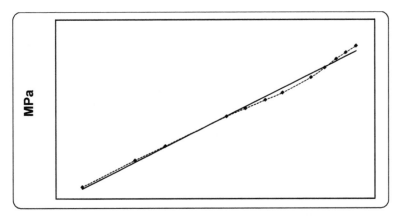

Figure 3-1. Plot Y (MPa) versus *X*, points reflect the field data.

This result is in good agreement with the value of $k = 13.6$ md obtained by the multiple-rate analysis technique (Earlougher 1977). From Eq. (3-18) we determine the value of the skin factor $s = -1.51$.

3.4.2 Simulated example

Test was designed by Schlumberger Well Services (Kucuk and Ayestaran 1985a,b). Reservoir rock and fluid data properties are given in Table 3-3.

Table 3-3. Reservoir rock and fluid data properties.

$h = 100$ ft	$\mu = 0.8$ cP
$\phi = 0.20$	$k = 40.0$ md (assumed)
$c_t = 0.000011$/psi	$r_w = 0.35$ ft
$B = 1.0$ RB/STB	$S = 10$ (assumed)

The wellbore flow rate and pressure for the first six (6) minutes of the test are presented in Table 3-4. The dimensionless flow rate was approximated by three polynomials with $n = 4$, $n = 5$, and $n = 6$. The coefficients a_0, a_1,...a_n are presented in Table 3-5. A multiple regression analysis computer program was used to estimate the values of these coefficients.

When a fourth-degree polynomial was applied, the values of s and k were estimated with low accuracy (Table 3-6). For fifth and sixth degree polynomials, computed and assumed values of the skin factor and formation permeability are compared in Table 3-6 and indicate a good match. The suggested technique for analyzing VFRD tests reduces the required test data. In this example, only pressure and flow rate data obtained during the first six minutes of the test were used.

Table 3-4. Synthetic drawdown pressure and flow data, Schlumberger well services (Kucuk and Ayestaran 1985a,b). Data presented by Dr. F. Kucuk. $q_{ref} = 175.480$ STB/D.

t, hrs	q_D	p_{wf}, psi	t, hrs	q_D	p_{wf}, psi
0.00	0.00	4400,0	0.055	8.5427	3813.3
0.010	0.5504	4365.6	0.060	9.4045	3751.2
0.015	1.1490	4325.2	0.064	10.0190	3706.8
0.020	1.9245	4273.1	0.070	10.9190	3640.8
0.025	2.7948	4214.5	0.074	11.4700	3600.6
0.030	3.7267	4150.5	0.080	12.1970	3546.2
0.035	4.7066	4083.5	0.086	12.8650	3496.3
0.040	5.7468	4011.0	0.090	13.2530	3466.6
0.045	6.7151	3943.5	0.094	13.6410	3437.1
0.050	7.6493	3876.6	0.100	14.1420	3398.2

Table 3-5. Coefficients in Eq. (3-7). Sigma is the sum of squared residuals.

Coefficient	$n = 4$	$n = 5$	$n = 6$
a_0, E-01	−0.177556	0.036666	0.007683
a_1, E+01	0.738242	−0.495642	−0.058943
a_2, E+04	0.580749	0.677438	0.626877
a_3, E+05	−0.709792	−0.976697	−0.765994
a_4, E+06	0.263579	0.566560	0.167502
a_5, E+07	-	0.120927	0.229248
a_6, E+08	-	-	−0.115644
Sigma	0.008875	0.002598	0.002278

Table 3-6. Interpretation results of the simulated VFR Test.

Parameter	$n = 4$	$n = 5$	$n = 6$
M	6.9137	5.4997	5.8113
Int	80.826	78.44I	78.969
k, md	33.016	41.505	39.279
S	7.704	10.551	9.803
$\Delta k/k$, %	17.460	3.762	1.802
$\Delta s/s$, %	22.960	5.510	1.970

Thus, a new method for analyzing variable flow rate drawdown tests is suggested. A Cartesian plot or a linear regression computer program can be used to obtain skin factor and formation permeability.

4

Short Term Testing Method for Stimulated Wells–Field Examples

4.1 STT Method: A Brief Review

A new Short Term Testing (STT) method is developed and presented in this chapter (Kutasov 2013). The STT method is based on a new semi-theoretical analytical solution of the diffusivity equation for a cylindrical source and is used to describe the transient buildup or drawdown pressures for stimulated wells. This solution is in good agreement with the results of a numerical ("Exact") solution of the diffusivity equation. The analytical solution allows to determine the bottom-hole pressure (or pressure derivative) at linear and linear-radial flow transition with high accuracy. For small values of time the STT method gives a half unit slope straight line (dimensionless pressure drop is proportional to the square root of time), and for large values of time the conventional formula for a radial flow can be obtained. The STT method can be used for any values of flowing or shut-in time. For this reason the principle of superposition can be used without any limitations. Below we demonstrate application of the STT method for two acidized oil wells.

Let us assume that a well is producing at a constant flow rate from an infinite-acting reservoir and: reservoir is a homogeneous and isotropic porous medium of uniform thickness, porosity and permeability are constant, fluid of small and constant compressibility, constant fluid viscosity, small pressure gradients, negligible gravity forces and laminar flow in the reservoir and the Darcy law can be applied. For a well as a cylindrical source it is necessary to obtain the solution of the diffusivity equation under following boundary and initial conditions:

$$p(t = 0, r) = p_i, \qquad r_w \leq r < \infty, \qquad t = 0, \tag{4-1}$$

$$\left(r \frac{\partial p}{\partial r} \right)_{r=r_w} = \frac{q\mu}{2\pi kh}, \qquad t > 0, \tag{4-2}$$

$$p(t, r \to \infty) \to p_i, \qquad t > 0, \tag{4-3}$$

where h is the reservoir thickness, r_w is the well radius, p_i is the initial reservoir pressure, q is the flow rate, k is the permeability, and μ is the fluid viscosity. It is well known, that in this case, the diffusivity equation has a solution in complex integral form (Van Everdingen and Hurst 1949; Carslaw and Jaeger 1959).

Chatas (Lee 1982, 106–107) tabulated this integral for a wide range of values of t_D. We will call the numerical solution by Chatas as an "Exact" solution (p_{DCh}). Below we will assume that for hydraulically fractured and acidized wells the apparent radius concept can be used. In practice this means that the radius of investigation is considerably larger than the effective fracture half-length. From section 4.6 (see Tables 4-8 and 4-9) follows that conventional approach (based on the Ei function) can be used only for developed radial fluid flow (at large values of dimensionless time, t_D). The suggested method can be used for any values of dimensionless time. The underlying assumptions in both mentioned methods are the same. The early-time skin is expressed through the apparent well radius.

4.2 Solutions for Cylindrical and Linear Sources

It is clear from physical considerations that, for large values of dimensionless time, the solutions for cylindrical and linear sources should converge. To develop the solution for a linear source, the boundary condition expressed by Eq. (4-2) should be replaced by the condition

$$\lim_{r\to 0}\left(r\frac{\partial p}{\partial r}\right)=\frac{q\mu}{2\pi kh},\qquad t>0,\tag{4-4}$$

and the well-known solution in oilfield units for the pressure (at $r=r_w$) for an infinitely long linear source with a constant flow rate in an infinite-acting reservoir is (Lee 1982):

$$p_i - p_{wf} = 141.2\frac{qB\mu}{kh}p_D\left(t_D\right),\tag{4-5}$$

where B is the reservoir volume factor.

The dimensionless time in oilfield units is

$$t_D = \frac{0.0002637\,kt}{\phi c_t \mu r_w^2},\tag{4-6}$$

$$p_D\left(t_D\right) = -\frac{1}{2}Ei\left(-\frac{1}{4t_D}\right),\tag{4-7}$$

where c_t is the total compressibility, Ei is the exponential integral, and p_D is the dimensionless pressure.

In Table 4-8 (see section 4-6) function $p_D{}^*(t_D) = p_D(t_D)$ (Eq. (4-7)) and the "Exact" solution of Chatas are compared. Thus we can make a conclusion that only at $t_D > 1000$ the Ei-function approximates the bottom-hole pressure with high accuracy. Introducing the pressure drop due to skin (s)

$$p_i - p_{wf} = 141.2 \frac{qB\mu}{kh} \left[p_D(t_D) + s \right], \tag{4-8}$$

where p_{wf} is the bottom-hole flowing pressure.

Let us assume that the apparent radius concept can be used

$$r_{wa} = r_w e^{-s}, \tag{4-9}$$

where r_{wa} is the effective wellbore radius.

Then

$$p_i - p_{wf} = 141.2 \frac{qB\mu}{kh} \left[p_D(t_{Da}) \right] \tag{4-10}$$

and the dimensionless flowing time based on the effective wellbore radius is

$$t_{Da} = \frac{0.0002637 kt}{\phi c_t \mu r_{wa}^2}. \tag{4-11}$$

The dimensionless bottom-hole pressure for a well with skin is

$$p_D(t_D) = -\frac{1}{2} Ei\left(-\frac{1}{4t_D} \right) + s. \tag{4-12}$$

In the case of a damaged well the last expression has a clear physical meaning: to maintain the flow rate an additional pressure drop is required. But this is not a case for a stimulated well. Indeed, under some conditions the value of $p_D(t_D)$ can be zero or even negative. It is obvious that the values of $p_D(t_D) \le 0$ do not have any physical meaning.

Thus in some cases (for small values of dimensionless time) a cylindrical source (wellbore) cannot be substituted by a linear source.

4.3 The Basic Equation

Earlier we obtained a semi analytical equation for the wall temperature for a cylindrical source with a constant heat flow rate (Kutasov 2003). Due to the similarity in Darcy's and Fourier's laws the same differential diffusivity equation describes the transient flow of incompressible fluid in porous medium and heat conduction in solids. As a result, a correspondence exists between the following parameters: volumetric flow rate, pressure gradient, mobility, hydraulic diffusivity coefficient and heat flow rate, temperature gradient, thermal conductivity and thermal diffusivity. Thus, it is reasonable to assume that similar to the techniques and data processing procedures of temperature well tests can be applied to pressure well tests.

In terms of pressure, mobility (permeability viscosity ratio: k/μ), fluid flow rate (q), porosity and total compressibility product (ϕc_t) this equation (in oilfield units) can be rewritten in the following form (see also section 4.6)

$$p_{wf}(t) = p(t, r_w) = p_i - \frac{q\mu}{2\pi k} \ln\left[1 + \left(c - \frac{1}{a + \sqrt{t_D}}\right)\sqrt{t_D}\right],$$
(4-13)

$$a = 2.7010505, \qquad c = 1.4986055.$$
(4-14)

Let's introduce the dimensionless wall pressure

$$p_D(t_D) = \frac{2\pi k h(p_i - p_w)}{q\mu},$$
(4-15)

$$p_D(t_D) = \ln\left[1 + \left(c - \frac{1}{a + \sqrt{t_D}}\right)\sqrt{t_D}\right].$$
(4-16)

Values of p_D calculated from Eq. (4-16) (Eq. (4-51) in section 4.6) and results of the numerical solution are compared in Table 4-9 (section 4.6). The agreement between values of p_D calculated by these two methods is very good. For this reason the principle of superposition can be used without any limitations.

4.4 Pressure Buildup Test

Using Eq. (4-16) and principle of superposition for a well producing at rate q until time t_p, and shut-in thereafter, we obtain working formulas for a pressure buildup test. Let Δt be the shut-in time, then

$$p_{wf} = p_i - m \ln\left[1 + \left(c - \frac{1}{a + \sqrt{t_{Dap}}}\right)\sqrt{t_{Dap}}\right], \qquad t = t_p,$$
(4-17)

$$p_{ws} = p_i - m\left\{\ln\left[1 + \left(c - \frac{1}{a + \sqrt{t_{Da}}}\right)\sqrt{t_{Da}}\right] - \ln\left[1 + \left(c - \frac{1}{a + \sqrt{t_{Das}}}\right)\sqrt{t_{Das}}\right]\right\}, \qquad t > t_p,$$
(4-18)

$$t_{Dap} = \frac{0.0002637 k t_p}{\phi c_t \mu r_{wa}^2}, \quad t_{Da} = t_{Dap}\, \beta, \quad t_{Das} = t_{Dap}\, \gamma,$$
(4-19)

$$\beta = \frac{t_p + \Delta t}{t_p}, \qquad \gamma = \frac{\Delta t}{t_p},$$
(4-20)

$$m = 141.2\frac{qB\mu}{kh}, \qquad k = 141.2\frac{qB\mu}{mh},$$
(4-21)

where Δt is the shut-in time.

It is easy to see that for large values of t_{Da} and t_{Das} we obtain the well-known Horner equation

$$p_{ws} = p_i - \frac{m}{2} \ln \frac{t_p + \Delta t}{\Delta t}. \qquad (4\text{-}22)$$

Let us consider two cases.

A. The value of the initial reservoir pressure is known. At least two measurements of pressure p_{wf} and p_{ws1} (at $t = t_p$ and $\Delta t = \Delta t_1$) are needed to calculate the formation permeability and skin factor.

Let

$$F\left(t_{Dap}\right) = \ln\left[1 + \left(c - \frac{1}{a + \sqrt{t_{Dap}}}\right)\sqrt{t_{Dap}}\right], \qquad (4\text{-}23)$$

$$\psi\left(t_{Dap},\beta\beta\right) = \ln\left[1 + \left(c - \frac{1}{a + \sqrt{t_{Dap}\beta}}\right)\sqrt{t_{Dap}\beta}\right] - \ln\left[1 + \left(c - \frac{1}{a + \sqrt{t_{Dap}\gamma}}\right)\sqrt{t_{Dap}\gamma}\right]. \qquad (4\text{-}24)$$

Then

$$p_{wf} = p_i - mF\left(t_{Dap}\right), \qquad (4\text{-}25)$$

$$p_{ws} = p_i - m\Psi\left(t_{Dap},\beta,\gamma\right) \qquad (4\text{-}26)$$

and

$$\theta = \frac{p_i \text{-} p_{fw}}{p_i - p_{ws1}} = \frac{F\left(t_{Dap}\right)}{\Psi\left(t_{Dap},\beta_1,\gamma_1\right)} = f\left(t_{Dap}\right), \qquad (4\text{-}27)$$

where Δt is the shut-in time

$$\beta_1 = \frac{t_p + \Delta t_1}{t_p}, \qquad \gamma_1 = \frac{\Delta t_1}{t_p}. \qquad (4\text{-}28)$$

Let us assume that the absolute accuracy of the ratio of θ is ε, then solving the following equation we calculate the value of t_{Dap}

$$\theta - f(t_{Dap}) = \varepsilon. \qquad (4\text{-}29)$$

The Newton method (Grossman 1977) was used for solving Eq. (4-29). In this method a solution of an equation is sought by defining a sequence of numbers which become successively closer and closer to the solution. The conditions, which guarantee that Newton method in our case will work and provide a unique solution, are satisfied (Grossman 1977). From Eq. (4-25) we can calculate the value of m. Then the formation permeability can be determined from Eq. (4-21) and, finally the skin factor can be estimated from Eq. (4-19).

$$r_{wa} = \sqrt{\frac{0.0002637kt_p}{t_{Dap}\phi c_t \mu}}, \qquad s = -\ln\frac{r_{wa}}{r_w}. \qquad (4\text{-}30)$$

B. The value of the initial reservoir pressure is not known. At least three measurements of pressure (p_{wf}, p_{ws1}, p_{ws2}) at $t = tp$, $\Delta t = \Delta t_1$, and $\Delta t = \Delta t_2$ are needed to calculate the formation permeability, initial reservoir pressure and skin factor. From Eqs. (4-23) and (4-24) we obtain

$$\theta = \frac{p_{wf}-p_{ws1}}{p_{wf}-p_{ws2}} = \frac{F(t_{Dap})-\Psi(t_{Dap},\beta_1,\gamma_1)}{F(t_{Dap})-\Psi(t_{Dap},\beta_2,\gamma_2)} = f(t_{Dap}), \tag{4-31}$$

$$\beta_1 = \frac{t_p+\Delta t_1}{t_p}, \qquad \gamma_1 = \frac{\Delta t_1}{t_p}, \qquad \beta_2 = \frac{t_p+\Delta t_2}{t_p}, \qquad \gamma_2 = \frac{\Delta t_2}{t_p}. \tag{4-32}$$

The procedure of solving the last equation and calculation of the value of t_{Dap} is similar to that of Eq. (4-29). From the following equation we can determine the value of m

$$P_{ws2} - P_{ws1} = m\left[\Psi\left(t_{Dap},\beta_1,\gamma_1\right)-\Psi\left(t_{Dap},\beta_2,\gamma_2\right)\right] \tag{4-33}$$

and from Eqs. (4-17), (4-21) and (4-30) we can determine values of the initial pressure, formation permeability, and skin factor. From pressure buildup test data a set of values p_i, k, and s is calculated and the average (or using statistical methods) values of these parameters are determined. To speed up the calculations after Eqs. (4-29) and (4-31) we prepared two computer programs. In the subroutine, which utilizes the Newton method, the following parameters were used: (a) the starting value of t_{Dap} is 0.01; (b) the time increment is 2; (c) the absolute accuracy of the ratio of θ is $\varepsilon = 0.001$.

4.5 Field Examples

Below we present two examples of application of the STT method to a pressure buildup tests conducted in two acid-fractured oil wells (McDonald 1983). We will use only early test data. We will also show that the utilized pressure-time data were recorded during the linear or linear-radial transition flow periods.

Oil Wells IS-21, PT-1 and South Dome Well IS-7, PT-1 were completed on the acid-fractured Shuaiba reservoir, Qatar (McDonald 1983). The reservoir, well and oil properties data are presented in Table 4-1.

Table 4-1. Reservoir and well data for two oil wells (McDonald 1983).

Well IS-21, PT-1.	$t_p = 263$ hours	$\mu = 1.5$ cP
	$p_i = 2303$ psi	$c_t = 10.4 \cdot 10^{-6}$ psi^{-1}
	$q = 870$ B/D	$B = 1.22$ RB/STB
	$h = 76$ ft	$r_w = 0.51$ ft
	$\phi = 0.21$	$p_{wf} = 1,850$ psi
Well IS-7, PT-1.	$t_p = 166$ hours	$\mu = 2.4$ cP
	$p_i = 2270$ psi	$c_t = 9.1 \cdot 10^{-6}$ psi^{-1}
	$q = 342$ B/D	$B = 1.18$ RB/STB
	$h = 101$ ft	$r_w = 0.51$ ft
	$\phi = 0.19$	$p_{wf} = 1,882$ psi

4.5.1 Oil well IS-21, PT-1

A pressure buildup test was conducted in this well; the duration of the shut-in period was 48 hours. Using the type-curve matching approach it was determined that the start of the Horner straight line to begin after approximately 30 hours shut-in time. As mentioned by McDonald (1983): "*Therefore, the straight line drawn through the late-time data on the Horner plot (Figure 4-1), although not well defined, still should be correct Horner straight line*". From the slope $m = 186$ psi/cycle the average permeability of 18.3 md was determined and by using the value of $p_{1\,hr} = 1,853$ psi the skin factor of –4.7 was calculated. Assuming that the values of skin factor (–4.7) and formation permeability (18.3 md) are correct ones, we tabulated Eq. (4-7), replacing t_D by t_{Da} (Table 4-2). From Table 4-2 follows that using Eq. (4-7) at $\Delta t = 30$ hours allows to calculate the dimensionless pressure with a relative accuracy of about 3.5%. Note also that the duration of the test is 311 (263 + 48) hours and value of t_{Da} (at 311 hrs) is less than 150 and the *Ei*-function describes pressure buildup with low accuracy (see Tables 4-8 and 4-9 in section 4.6). We should also to mention that deviations $p_{DCh} - p_D{}^*$ can be wrongly attributed to the wellbore storage effect. The Reader can also see that the method of pressure derivatives cannot be used if the *Ei* solution is erroneously utilized.

First we consider a case when the initial pressure is known. In Table 4-3 we present results of calculation of formation permeability and skin factor. Only pressure records taken for the first 16 hours of shut-in period were used. Comparison between the values of skin factor $s = -4.7$ and formations permeability $k = 18.3$ md obtained from the Horner plot (large shut-in times) and those ($s = -4.80$, $k = 17.25$ md) determined by using the suggested method (using only the early time data) shows that these parameters are in good agreement.

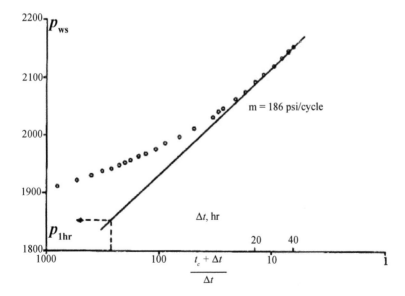

Figure 4-1. Pressure-buildup data for Well IS-21, PT-1; Horner plot (McDonald 1983). Courtesy of Society of Petroleum Engineers.

Table 4-2. Comparison of the values of dimensionless pressure for the Well IS-21, PT-1, t_{Da} (1 hour) = 0.469, p_{DCh} is the "Exact" solution, $p_D{}^*$—Eq. (4-7), $R = (p_{DCh} - p_D{}^*)/p_{DCh} \cdot 100$.

t_{Da}	Time, hr	p_{DCh}	$p_D{}^*$	$p_{DCh} - p_D{}^*$	R, %
1.0	2.13	0.8019	0.5221	0.2798	34.89
2.0	4.26	1.0195	0.8117	0.2078	20.38
3.0	6.40	1.1665	0.9947	0.1718	14.73
4.0	8.53	1.2750	1.1285	0.1465	11.49
5.0	10.66	1.3625	1.2339	0.1286	9.44
6.0	12.79	1.4362	1.3210	0.1152	8.02
7.0	14.93	1.4997	1.3952	0.1045	6.97
8.0	17.06	1.5557	1.4598	0.0959	6.17
9.0	19.19	1.6057	1.5169	0.0888	5.53
10.0	21.32	1.6509	1.5683	0.0826	5.01
15.0	31.98	1.8294	1.7669	0.0625	3.42
20.0	42.64	1.9601	1.9086	0.0515	2.63
30.0	63.97	2.1470	2.1093	0.0377	1.76
40.0	85.29	2.2824	2.2521	0.0303	1.33
50.0	106.61	2.3884	2.3630	0.0254	1.06
60.0	127.93	2.4758	2.4538	0.0220	0.89
70.0	149.25	2.5501	2.5306	0.0195	0.77
80.0	170.58	2.6147	2.5971	0.0176	0.67
90.0	191.90	2.6718	2.6558	0.0160	0.60
100.0	213.22	2.7233	2.708	0.0149	0.55
150.0	319.83	2.9212	2.9107	0.0105	0.36

Table 4-3. Processing of buildup data by the STT method, Oil Well IS-21, PT-1, p_i = 2303 psi. Average values: k = 17.25 md, s = –4.80.

Δt, min	p_{ws}, psi	k, md	s	t_{Dap}
60	1942	17.81	–4.784	101.3
70	1949	17.99	–4.761	107.1
80	1953	17.80	–4.785	101.1
90	1957	17.69	–4.800	97.6
105	1963	17.62	–4.808	95.5
120	1967	17.38	–4.840	88.4
150	1977	17.36	–4.843	87.8
180	1986	17.39	–4.838	88.8
240	1998	17.08	–4.878	80.5
330	2013	16.83	–4.910	74.4
480	2032	16.55	–4.946	68.1
540	2041	16.79	–4.916	73.4
600	2047	16.74	–4.922	72.4
780	2063	16.71	–4.925	71.7
960	2077	16.81	–4.913	73.9

The fracture length was defined by

$$L = 2r_{wa}.$$

Now let us consider a more general case when the value of p_i is not known. The input data and results of calculations are presented in Table 4-4. The average value of formation permeability is 19.35 md and the average value of the skin factor is –4.76. In this case also the values of s and k are very close to the parameters obtained from the Horner plot. The STT method can be also used for evaluation the efficiency of wellbore stimulation.

Table 4-4. Determination of formation permeability, skin factor, and initial pressure by the STT method, Oil Well IS-21, PT-1 $p_{wf}=$ 1,850 psi. Average values: $k=$ 19.35 md, $s=$ –4.76.

Δt_1, hrs	Δt_2, hrs	p_{ws1}, psi	p_{ws2}, psi	k, md	s	p_i, psi
2	4	1967	1998	21.31	–4.667	2257
2	9	1967	2041	20.19	–4.715	2269
2	10	1967	2047	20.11	–4.718	2269
2	13	1967	2063	19.61	–4.740	2275
2	16	1967	2077	18.95	–4.769	2283
3	9	1986	2041	21.17	–4.659	2260
3	10	1986	2047	20.96	–4.668	2262
3	13	1986	2063	20.17	–4.705	2270
3	16	1986	2077	19.31	–4.745	2280
4	9	1998	2041	19.44	–4.758	2275
4	10	1998	2047	19.40	–4.760	2276
4	13	1998	2063	18.86	–4.787	2282
4	13	1998	2063	18.86	–4.787	2282
4	16	1998	2077	18.10	–4.825	2290
9	16	2041	2077	16.58	–4.929	2305
10	16	2047	2077	16.07	–4.963	2311

4.5.2 Oil well IS-7, PT-1

A pressure buildup test was conducted in this well; the duration of the shut-in period was 59 hours. Using the type-curve matching approach it was determined that the start of the Horner straight line to begin after approximately 15 hours shut-in time. From the slope $m =$ 143 psi/cycle (Figure 4-2) the average permeability of 10.9 md was determined and by using the value of $p_{1hr}=$ 1,952 psi the skin factor of –3.8 was calculated.

Assuming that the values of skin factor (–3.8) and formation permeability (10.9 md) are correct ones, we tabulated Eq. (4-7), replacing t_D by t_{Da} (Table 4-5). From Table 4-5 follows that using Eq. (4-7) at $\Delta t =$ 15 hours allows to calculate the dimensionless pressure with a relative accuracy of about 2.6%. Note also that the duration of shut-in period of the test is 59 hours and value of t_{Das} (59 hrs) is less than 80 and the Ei-function describes pressure buildup with low accuracy (see Tables 4-5 and 4-8).

Consider a case when the initial pressure is known. In Table 4-6 we present results of calculation of formation permeability and skin factor. Only pressure records taken for the first 12 hours of shut-in period were used. It was estimated that the average values are: $k = 11.13$ md and $s = -3.73$.

Now let consider a more general case when the value of p_i is not known. The input data and results of calculations are presented in Table 4-7. The average value of formation permeability is 13.21 md and the average value of the skin factor is –3.52.

Thus the estimated values of s and k in both cases and those ($s = -3.8$, $k = 10.9$ md) obtained from the conventional Horner plot are in satisfactory agreement. Wellbore storage has been recognized as affecting short-time pressure behavior (Kutasov 2003).

In our cases the storage effect is not significant at $\Delta t > 1$ hr (Well IS-21, PT-1) and at $\Delta t > 55$ min (Well IS-7, PT-1). Tables 4-3 and 4-6 indirectly confirm the above mentioned statement. Indeed, at these times pressure behavior approximates by the suggested equation (i.e., which governs the linear and transitional fluid flow). Well storage shows (Earlougher 1977) (in the cases when Horner plot can be used and the *Ei* solution is applicable), a unit slope straight line in a log-log plot of Δp ($p_i - p_s$) versus t_s. The end of this plot determines the duration of wellbore storage. The suggested formula at large values of dimensionless time transforms to the conventional Horner plot (Eq. (4-22)).

Tables 4-2 and 4-5 show what is the accuracy of determination p_D when the conventional solution (based on the *Ei* function) for various time intervals instead of the suggested equation is used. For example if the time interval is $2.13 \leq t \leq 17.06$ hours (first field example) the relative accuracy of p_D is between 6.2 and 34.9% (Table 4-2). Similarly for the second field example if the time interval is $3.0 \leq t \leq 30.0$ hours the relative accuracy of p_D is between 1.3 and 11.5% (Table 4-5).

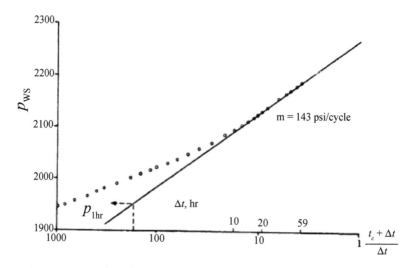

Figure 4-2. Pressure-buildup data for Well IS-7, PT-1, Horner plot. Courtesy of Society of Petroleum Engineers.

Table 4-5. Comparison of the values of dimensionless pressure for the Well IS-7, PT-1, t_{Da} (1 hour) = 1.333, p_{DCh} is the "Exact" solution, p_D^* is obtained from Eq. (4-7), $R = \dfrac{p_{Dch} - p_D^*}{p_{Dch}} \cdot 100\%$.

t_{Da}	t, hrs	p_{DCh}	p_D^*	$p_{Dch} - p_D^*$	R, %
0.10	0.08	0.3144	0.0125	0.3019	96.04
0.20	0.15	0.4241	0.0732	0.3509	82.74
0.30	0.23	0.5024	0.1463	0.3561	70.88
0.40	0.30	0.5645	0.2161	0.3484	61.71
0.50	0.38	0.6167	0.2799	0.3368	54.62
0.60	0.45	0.6622	0.3376	0.3246	49.02
0.70	0.53	0.7024	0.3900	0.3124	44.47
0.80	0.60	0.7387	0.4378	0.3009	40.73
0.90	0.68	0.7716	0.4817	0.2899	37.58
1.00	0.75	0.8019	0.5221	0.2798	34.89
2.00	1.50	1.0195	0.8117	0.2078	20.38
3.00	2.25	1.1665	0.9947	0.1718	14.73
4.00	3.00	1.2750	1.1285	0.1465	11.49
5.00	3.75	1.3625	1.2339	0.1286	9.44
6.00	4.50	1.4362	1.3210	0.1152	8.02
7.00	5.25	1.4997	1.3952	0.1045	6.97
8.00	6.00	1.5557	1.4598	0.0959	6.17
9.00	6.75	1.6057	1.5169	0.0888	5.53
10.00	7.50	1.6509	1.5683	0.0826	5.01
15.00	11.25	1.8294	1.7669	0.0625	3.42
20.00	15.00	1.9601	1.9086	0.0515	2.63
30.00	22.51	2.1470	2.1093	0.0377	1.76
40.00	30.01	2.2824	2.2521	0.0303	1.33
50.00	37.51	2.3884	2.3630	0.0254	1.06
60.00	45.01	2.4758	2.4538	0.0220	0.89
70.00	52.51	2.5501	2.5306	0.0195	0.77
80.00	60.02	2.6147	2.5971	0.0176	0.67
90.00	67.52	2.6718	2.6558	0.0160	0.60
100.00	75.02	2.7233	2.7084	0.0149	0.55
150.00	112.53	2.9212	2.9107	0.0105	0.36
200.00	150.04	3.0636	3.0543	0.0093	0.30
250.00	187.55	3.1726	3.1658	0.0068	0.22
300.00	225.06	3.2630	3.2568	0.0062	0.19

Table 4-6. Processing of the buildup data by the STT method, Well IS-7, PT-1, $p_i = 2270$ psi. Average values: $k = 11.13$ md, $s = -3.73$.

Δt, min	P_{ws}, psi	k, md	s	t_{Dap}
55	2002	11.88	−3.546	401.0
70	2010	11.75	−3.578	371.9
85	2017	11.67	−3.598	354.6
100	2021	11.45	−3.653	311.5
130	2030	11.25	−3.701	278.5
160	2038	11.14	−3.728	261.4
205	2047	10.95	−3.776	233.0
265	2058	10.84	−3.803	218.8
355	2070	10.64	−3.851	195.0
475	2085	10.65	−3.849	195.8
600	2098	10.72	−3.832	204.0
720	2105	10.50	−3.884	180.3

Table 4-7. Determination of formation permeability, skin factor, and initial pressure by the STT method for Oil Well IS-7, PT-1. Average values: $k = 13.21$ md, $s = -3.52$.

Δt_1, min	Δt_2, min	P_{ws1}, psi	P_{ws2}, psi	k, md	s	P_i, psi
100	355	2021	2070	15.37	−3.30	2215
100	475	2021	2085	14.24	−3.40	2228
100	600	2021	2098	13.44	−3.47	2239
100	720	2021	2105	13.68	−3.45	2236
130	355	2030	2070	14.90	−3.35	2219
130	475	2030	2085	13.72	−3.46	2233
130	600	2030	2098	12.91	−3.54	2244
130	720	2030	2105	13.24	−3.51	2240
160	475	2038	2085	13.49	−3.49	2236
160	600	2038	2098	12.63	−3.58	2247
160	720	2038	2105	13.03	−3.54	2242
205	475	2047	2085	12.70	−3.59	2244
205	600	2047	2098	11.89	−3.68	2255
205	720	2047	2105	12.45	−3.62	2247
265	600	2058	2098	11.47	−3.73	2260
265	720	2058	2105	12.23	−3.65	2250

We can conclude that a new Short Term Testing (STT) method for stimulated wells is developed. The STT method is based on a new semi-theoretical analytical solution of the diffusivity equation for a cylindrical source and is used to describe the transient buildup or drawdown pressures for stimulated wells. This suggested solution gives a good agreement with the results of a numerical solution of the diffusivity equation. The analytical solution allows to determine the bottom-hole pressure (or pressure derivative) at linear flow and at linear-radial flow transition with high accuracy. The STT method can be used for any values of flowing or shut-in time. For this reason the principle of superposition can be used without any limitations. The field data processing technique is validated by a buildup test conducted in two acid-fractured oil wells. As is shown in section 4.6 (see Tables 4-8 and 4-9) the suggested methods can be used for any values of dimensionless time, and consequently for any a range of permeabilities/mobilities. The method will work best at large values of negative skin factors (i.e., for stimulated wells).

4.6 Derivation of Equation (4-13)

We will begin with the comparison of solutions of the diffusivity equation for cylindrical and linear sources. Let us assume that a well is producing at a constant flow rate from an infinite-acting reservoir and: reservoir is a homogeneous and isotropic porous medium of uniform thickness, porosity and permeability are constant, fluid of small and constant compressibility, constant fluid viscosity, small pressure gradients, negligible gravity forces and laminar flow in the reservoir and the Darcy law can be applied. For a well as a cylindrical source it is necessary to obtain the solution of the diffusivity equation under following boundary and initial conditions:

(1) $p(t=0,r) = p_i, \quad r \geq r_w,$

(2) $\left(r \dfrac{\partial p}{\partial r} \right)_{r_w} = \dfrac{q\mu}{2\pi kh}, \quad t > 0,$

(3) $p(t, r \to \infty) = p_i, \quad t > 0,$

where h is the reservoir thickness, r_w is the well radius, p_i is the initial reservoir pressure, q is the flow rate, k is the permeability, and μ is the fluid viscosity.

The solution (at $r = r_w$) is (Carslaw and Jaeger 1959):

$$p_D(t_D) = \frac{4}{\pi^2} \int_0^\infty \frac{1 - e^{-u^2 t}}{u^3 \left[J_1^2(u) + Y_1^2(u) \right]} du, \tag{4-34}$$

where $J_1(u)$ is the Bessel function of the first kind and $Y_1(u)$ is the modified Bessel function of the first kind. Chatas (Lee 1982, 106–107) tabulated this integral over a wide range of values of t_D (Table 4-8, function p_{DCh}). Below we will call the function p_{DCh} as an "Exact" solution.

The dimensionless time (t_D) in oilfield units is

$$t_D = \frac{0.0002637 kt}{\phi c_t \mu r_w^2}, \tag{4-35}$$

where ϕ is porosity, and c_t is the total compressibility.

To develop the solution for a linear source the second condition is replaced by the condition

$$(2^*) \quad \lim_{r \to 0} \left(r \frac{\partial p}{\partial r} \right) = \frac{q\mu}{2\pi kh}, \quad t > 0$$

and the well-known solution for the pressure at a well producing at a constant rate in an infinite-acting reservoir is

$$p_i - p_{wf} = 141.2 \frac{qB\mu}{kh} p_D(t_D), \tag{4-36}$$

where B is the reservoir volume factor.

$$p_D(t_D) = -\frac{1}{2} Ei\left(-\frac{1}{4t_D}\right). \tag{4-37}$$

In Table 4-8 the function $p_{D^*}(t_D) = p_D(t_D)$ (Eq. (4-37)) and the "Exact" solution are compared.

Table 4-8. Comparison of the values of dimensionless pressure, p_{DCh} is the "Exact" solution, p_{D^*} - Eq. (4-37), $R = \frac{p_D^* - p_{Dch}}{p_{Dch}} \cdot 100\%$.

t_D	p_{DCh}	p_D^*	$p_D^* - p_{DCh}$	$R, \%$
0.1	0.3144	0.0125	−0.3019	−96.04
0.4	0.5645	0.2161	−0.3484	−61.71
0.8	0.7387	0.4378	−0.3009	−40.73
1.0	0.8019	0.5221	−0.2798	−34.89
1.4	0.9160	0.6582	−0.2578	−28.14
2	1.0195	0.8117	−0.2078	−20.38
4	1.2750	1.1285	−0.1465	−11.49
6	1.4362	1.3210	−0.1152	−8.02
8	1.5557	1.4598	−0.0959	−6.17
10	1.6509	1.5683	−0.0826	−5.01
15	1.8294	1.7669	−0.0625	−3.42
20	1.9601	1.9086	−0.0515	−2.63
30	2.1470	2.1093	−0.0377	−1.76
40	2.2824	2.2521	−0.0303	−1.33
60	2.4758	2.4538	−0.0220	−0.89
80	2.6147	2.5971	−0.0176	−0.67
100	2.7233	2.7084	−0.0149	−0.55
200	3.0636	3.0543	−0.0093	−0.30
500	3.5164	3.5121	−0.0043	−0.12
800	3.7505	3.7470	−0.0035	−0.09
950	3.8355	3.8329	−0.0026	−0.07
1000	3.8584	3.8585	0.0001	0.00

Thus we can make a conclusion that only at $t_D > 1000$ the *Ei*-function approximates the bottom-hole pressure with high accuracy. Introducing the pressure drop due to skin

$$p_i - p_{wf} = 141.2 \frac{qB\mu}{kh} \Big[p_D(t_D) + s \Big].$$

(4-38)

Let us assume that the apparent radius concept can be used

$$r_{wa} = r_w e^{-s}.$$

(4-39)

Then

$$p_i - p_{wf} = 141.2 \frac{qB\mu}{kh} \Big[p_D(t_{Da}) \Big], \quad t_{Da} = \frac{0.0002637kt}{\phi\mu ctr_{wa}^2}.$$

(4-40)

From Table 4-8 for $t_D > 1000$ one can observe the convergence of solutions for cylindrical and linear sources. Then for values of $t_D \to \infty$ the logarithmic approximation of the *Ei* function can be used, and

$$p_D(t_D) = -\frac{1}{2} Ei\left(-\frac{1}{4t_D}\right) = \frac{1}{2}(\ln t_D + \ln 4 - 0.57722) = \frac{1}{2}(\ln t_D + 0.80907).$$

(4-41)

where 0.57722 is the Euler's constant. The last formula for values of $t_D \to \infty$ can be transformed to

$$p_D(t_D) = \frac{1}{2}(\ln t_D + 0.80907) = \ln \sqrt{t_D e^{0.80907}} = \ln\left(1 + c\sqrt{t_D}\right),$$

(4-42)

$$c = \sqrt{e^{0.80907}}, \qquad c = 1.4986055.$$

(4-43)

It is known (Carslow and Jaeger 1959) that for an infinite cylindrical source at values of $t_D \to 0$ a linear flow exist and the dimensionless wall temperature is

$$p_D(t_D) = 2\sqrt{\frac{t_D}{\pi}}$$

(4-44)

From Eq. (4-44) follows that for a linear regime of flow a half-unit slope straight line exists in a log-log plot of p_D versus t_D. The duration of the linear regime of flow is very short. For example, for $t_D = 0.005$ the value of $p_D = 0.0794$ (Eq. (4-44)).

And this agrees well with the "exact" value of 0.0774 (Table 4-9). Thus time interval for the linear regime of flow is $0.00 \le t_D \le 0.005$; for transitional regime the time interval is $0.005 \le t_D \le 1000$; and a developed radial regime of flow exists at $t_D > 1000$ (Table 4-9).

$$p_D(t_D) = \ln\left(1 + b\sqrt{t_D}\right), \qquad b = \frac{2}{\sqrt{\pi}}, \qquad b = 1.128379.$$

(4-45)

By using the above presented relationships, and applying the trial and error method to approximate Chatas values of p_D, we have found that for any values of dimensionless time the following semi theoretical formula is a good approximation of the dimensionless bottom-hole pressure (Kutasov 2003).

$$p_D(t_D) = \ln\left(1 + D\sqrt{t_D}\right),$$

(4-46)

$$D = c - \frac{1}{a + \sqrt{t_D}}.$$

(4-47)

Using the value of $D(t_D \rightarrow 0) = b$ we can determine the coefficient a:

$$\frac{1}{a} = \sqrt{e^{0.80907}} - \frac{2}{\sqrt{\pi}}, \qquad a = 2.7010505$$

(4-48)

and

$$p(t, r_w) = p_i - \frac{q_h \mu}{2\pi k h} \ln\left[1 + \left(c - \frac{1}{a + \sqrt{t_D}}\right)\sqrt{t_D}\right].$$

(4-49)

Thus, we obtained a semi-theoretical Eq. (4-13) of the bottom-hole pressure $p_{wf} = p(t, r_w)$ for a well produced at a constant flow rate. Introducing the dimensionless bottom-hole pressure

$$p_D(t_D) = \frac{2\pi k h(p_i - p_w)}{q_h \mu}.$$

(4-50)

We obtain

$$p_D(t_D) = \ln\left[1 + \left(c - \frac{1}{a + \sqrt{t_D}}\right)\sqrt{t_D}\right].$$

(4-51)

Values of p_D calculated from Eq. (4-51) and results of the numerical solution are compared in Table 4-9. The agreement between values of p_D calculated by these two methods is very good. For this reason the principle of superposition can be used without any limitations.

Table 4-9. Comparison of the values of dimensionless pressure, p_{DCh} is the "Exact" solution, p_D obtained from Eq. (4-51), $R = \dfrac{p_D - p_{DCh}}{p_{DCh}} \cdot 100\%$.

t_D	p_{DCh}	p_D	$p_D - p_{DCh}$	R, %
0.0005	0.0250	0.0250	0.0000	−0.06
0.001	0.0352	0.0352	0.0000	−0.02
0.002	0.0495	0.0495	0.0000	−0.03
0.005	0.0774	0.0774	0.0000	−0.02
0.010	0.1081	0.1081	0.0000	0.00
0.020	0.1503	0.1503	0.0000	0.00
0.05	0.2301	0.2300	−0.0001	−0.03
0.10	0.3144	0.3141	−0.0003	−0.08
0.20	0.4241	0.4241	0.0000	−0.01
0.50	0.6167	0.6164	−0.0003	−0.05
1.00	0.8019	0.8013	−0.0006	−0.08
2.00	1.0195	1.0209	0.0014	0.14
5	1.3625	1.3605	−0.0020	−0.15
10	1.6509	1.6486	−0.0023	−0.14
20	1.9601	1.9571	−0.0030	−0.15
50	2.3884	2.3863	−0.0021	−0.09
100	2.7233	2.7212	−0.0021	−0.08
200	3.0636	3.0612	−0.0024	−0.08
500	3.5164	3.5150	−0.0014	−0.04
1000	3.8584	3.8601	0.0017	0.04

5

Determination of the Skin Factor for a Well Produced at a Constant Bottom-Hole Pressure

A new technique has been developed for analyzing the constant bottom-hole pressure (BHP) test data. The method presented in this chapter enables to calculate the skin factor for damaged and stimulated oil wells. Advantages of the BHP test are: (1) they do not require that the well be shut in; (2) the fluid production can be easily controlled (at constant flow rate tests the BHP is changing with time); and (3) wellbore storage effects on the test data are short-lived. It is assumed that the instantaneous flow rate and time data are available from a well produced against a constant bottom-hole pressure. Only records of the flowing time and flow rate data are required to compute the value of the skin factor.

A semi theoretical equation is used to approximate the dimensionless flow rate. This formula is used to obtain a quadratic equation for determining the skin factor. The accuracy of the basic equation will be shown below. An example of calculation is presented.

Reservoir waters associated with oil bearing formations contain a variety of dissolved solids (scales). At simultaneous production of oil and water precipitation of scales occurs. In vicinity of the well the pressure logarithmically increases with the radius. As a result the deposition of scales mainly occurs near the wellbore. Due to partial blocking of pores a skin is created. The size of the skin increases with the production time. A comprehensive study of essential mechanisms of formation damage by pore blocking is presented in the literature (Schechter 1992). Deposition of scales in the pores significantly reduces the permeability of formations around the wellbore. This results in pressure loss due to skin and corresponding reduction in well productivity. Well stimulation through acidicing or hydraulic fracturing may improve the well productivity even above the production levels corresponding to undamaged conditions. Quantitatively the skin is defined as a parameter (skin factor) which is dependent on the thickness and the effective permeability of the skin zone:

$$s = \left(\frac{k}{k_s} - 1\right) \ln \frac{r_s}{r_w}, \tag{5-1}$$

where k is the formation permeability, k_s is the effective permeability of the skin zone, r_w is the well radius, r_s is the skin radius and s is the skin factor.

It is more convenient to express the skin factor through the apparent (effective) well radius (Earlougher 1977):

$$r_{wa} = r_w e^{-s}. \tag{5-2}$$

For hydraulically fractured wells the r_{wa}/r_w ratio can be very large. We will assume that for stimulated wells ($s < 0$) or damaged wells ($s > 0$) the apparent radius concept can be used (Uraiet and Raghavan 1980). In practice this means that the radius of investigations should be considerably larger than the apparent radius.

The objective of this chapter is to determine the skin factor and estimate the effect of wellbore damage on the long term well productivity. For this reason we will assume that the formation permeability was estimated from a previous buildup or drawdown pressure well test. We should note that the basic Eq. (5-5) was used earlier to develop a step-pressure test which enables to determine formation permeability and skin factor (Kutasov 1987b).

5.1 Dimensionless Flow Rate

Let us assume that the well is producing against a constant bottom-hole pressure (BHP) from an infinite-acting reservoir. We will consider here only wells which are completed (perforated) across the entire producing interval. In this case the relationship between the flow rate and time for a well with a constant BHP in oil field units is (Lee 1982):

$$q = \frac{kh\left(p_i - p_{wf}\right)}{141.2 B\mu} q_D, \tag{5-3}$$

$$t_D = \frac{0.0002637 kt}{\phi c_t \mu r_{wa}^2}, \tag{5-4}$$

where q_D is dimensionless flow rate, c_t is the total compressibility, h is the reservoir thickness, μ is the viscosity, ϕ is the porosity, k is the formation permeability, p_i is the initial reservoir pressure, p_{wf} is the bottom-hole flowing pressure and t_D is the dimensionless time based on the apparent well bore radius r_{wa}. We should also note that Eq. (5-3) is widely used in the petroleum industry to forecast oil flow rates. Analytical expressions for the function $q_D = f(t_D)$ are available only for asymptotic cases or for large values of t_D (Jacob and Lohman 1952; Ehlig-Economides and Ramey 1981). The dimensionless flow rate was first calculated (in tabulated form) by Jacob and Lohman (1952). They computed the dimensionless flow rate from an analytic solution obtained by Smith (1937), using numerical integration. Sengul used the same solution for a wider range of dimensionless time and with more table entries. The numerical Laplace inversion of the solution to dimensionless production rate in Laplace space was used by Sengul (1983). Applying the trial and error method to approximate Sengul's values of dimensionless flow rate we have found (Kutasov 1987, 1999) that for any values of dimensionless production time semi-theoretical Eq. (5-5) can be used to forecast the dimensionless flow rate

$$q_D = \frac{1}{\ln\left(1+D\sqrt{t_D}\right)}, \qquad (5\text{-}5)$$

$$D = d + \frac{1}{\sqrt{t_D}+b}, \qquad (5\text{-}6)$$

$$d = \frac{\pi}{2}, b = \frac{2}{2\sqrt{\pi}-\pi}. \qquad (5\text{-}7)$$

In Table 5-1 values of q_D calculated after Eq. (5-5) and the results of a numerical solution (Sengul 1983) (q_D^*) are compared. The agreement between values of q_D and q_D^* calculated by these two methods is seen to be good.

Table 5-1. Comparison of values of dimensionless flow rate for a well with constant BHP: q_D^* is the numerical solution (Sengul 1983), q_D is obtained from Eq. (5-5).

t_D	q_D^*	q_D	$\Delta q/q^* \cdot 100\%$
0.0001	56.918	56.930	0.02
0.0002	40.392	40.405	0.03
0.0005	25.728	25.741	0.05
0.001	18.337	18.350	0.07
0.002	13.110	13.122	0.09
0.005	8.4694	8.4818	0.15
0.01	6.1289	6.1410	0.20
0.02	4.4716	4.4835	0.27
0.05	2.9966	3.0079	0.38
0.1	2.2488	2.2596	0.48
0.2	1.7152	1.7255	0.60
0.5	1.2336	1.2430	0.77
1	0.98377	0.99260	0.90
2	0.80058	0.80877	1.02
5	0.62818	0.63555	1.17
10	0.53392	0.54068	1.27
20	0.46114	0.46730	1.34
50	0.38818	0.39351	1.37
100	0.34556	0.35025	1.36
200	0.31080	0.31484	1.30
500	0.27381	0.27706	1.19
1,000	0.25096	0.25366	1.08
2,000	0.23151	0.23372	0.95
5,000	0.20986	0.21153	0.80
10,000	0.19593	0.19727	0.69
20,000	0.18370	0.18477	0.58
50,000	0.16966	0.17044	0.46
100,000	0.16037	0.16098	0.38

5.2 Skin Factor

Let

$$c = \exp\left(\frac{1}{q_D}\right) - 1, \; x = \sqrt{t_D},$$

(5-8)

then after simple transformations we obtain from Eq. (5-5)

$$x^2 + a_1 x + a_2 = 0 \qquad (5-9)$$

and

$$x = \sqrt{t_D} = -\frac{a_1}{2} + \sqrt{\frac{a_1^2}{4} - a_2}, \qquad (5-10)$$

where

$$a_1 = \frac{db - c + 1}{d}, \; a_2 = -\frac{bc}{d}. \qquad (5-11)$$

The apparent (effective) well radius is calculated from Eq. (5-4)

$$r_{wa} = \sqrt{\frac{0.0002637 kt}{t_D \phi c_t \mu}} \qquad (5-12)$$

and finally from Eq. (5-2)

$$s = -\ln \frac{r_{wa}}{r_w}. \qquad (5-13)$$

Thus, we obtained a quadratic equation for estimating the skin factor. In the practice the values of q_D are estimated from Eq. (5-3).

5.3 Example

Using the table (Sengul 1983) of $q_D^* = f(t_D)$ we generated data for this simulated example. The reservoir, well, and fluid data (Table 5-2) were chosen to allow to avoid interpolation of q_D^* values. Let us now assume that the well started to produce oil and the well's skin factor is equal to zero. Later, after some time of production, a constant BHP test was conducted and it was found that due to formation damage the skin

Table 5-2. Reservoir, well, and fluid data for the simulated example, B is the oil formation volume factor.

Assumed parameters	
$h = 50$ ft	$k = 34.84$ md
$\phi = 0.15$	$p_i = 3000$ psi
$c_t = 25 \cdot 10^{-6}$ 1/psi	$p_{wf} = 2500$ psi
$B = 1.028$, RB/STB	$\mu = 2.0$ cP
$r_w = 0.35$ ft	

factor increased to $s = 4.605$. To improve the well performance well stimulation was conducted and a new constant BHP test has shown that the skin factor was reduced to $s = -4.605$. The results of calculations derived from Eqs. (5-2)–(5-3) are presented in Table 5-3 and Figures 5-1 and 5-2. The results of calculations show that Eq. (5-4) can be used to compute the flow rate and skin factor. Indeed, the assumed and calculated values of the skin factor are in a very good agreement.

Table 5-3. Calculated values of the skin factor. For stimulated well t_D ($t = 1$ hr) = 1.00; for well with $s = 0$ t_D ($t = 1$ hr) = $1.0 \cdot 10^4$, and for a well with $s = 4.605$ t_D ($t = 1$ hr) = $1.0 \cdot 10^8$ (Kutasov and Kagan 2003b).

t, hrs	$q_D^* s = -4.605$		$q_D^* s = 0.0$		$q_D^* s = 4.605$	
0.1	2.24880	–4.599	0.25096	0.044	0.117420	4.611
0.2	1.71520	–4.597	0.23151	0.042	0.112860	4.608
0.5	1.23360	–4.594	0.20986	0.038	0.107340	4.606
1.0	0.98377	–4.591	0.19593	0.035	0.103510	4.604
1.5	0.86994	–4.589	0.18859	0.033	0.101390	4.603
2.0	0.80058	–4.587	0.18370	0.032	0.099942	4.602
3.0	0.71620	–4.585	0.17722	0.030	0.097966	4.602
4.0	0.66440	–4.583	0.17288	0.028	0.096611	4.601
5.0	0.62818	–4.581	0.16966	0.027	0.095585	4.600
6.0	0.60088	–4.580	0.16711	0.026	0.094763	4.600
7.0	0.57928	–4.579	0.16502	0.026	0.094079	4.600
8.0	0.56157	–4.578	0.16325	0.025	0.093494	4.599
9.0	0.54668	–4.577	0.16171	0.024	0.092984	4.599
10.0	0.53392	–4.577	0.16037	0.024	0.092533	4.599

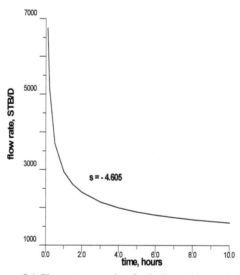

Figure 5-1. Flow rate versus time for the Example, $s = -4.605$.

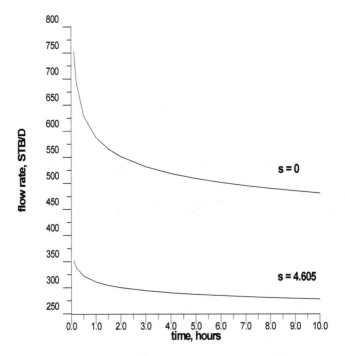

Figure 5-2. Flow rate versus time for the example, $s = 0$, and $s = 4.605$.

We can conclude that a new method has been developed for analyzing the constant bottom-hole pressure test data. The suggested method allows one to calculate the skin factor for damaged or stimulated wells.

6

Evaluation of the Efficiency of Wellbore Stimulation Operations

As a consequence of drilling and completion operations most wells have reduced permeability near the borehole (skin zone). In the oil/water production the radius of the skin zone increases with time. It is known that reservoir waters associated with oil-bearing formations contain a variety of dissolved solids (scales). As was mentioned in Chapter 5, deposition of scales in the pores significantly reduces the permeability of formations around the wellbore. This results in pressure loss through the skin and a corresponding reduction in well productivity. The objective of well stimulation through acidacion or hydraulic fracturing is to reduce the skin factor. The results of well stimulation operations can be expressed through skin factor (Eq. 6-1) or apparent well radius (Eq. 6-2).

$$s = \left(\frac{k}{k_s} - 1 \right) \ln \frac{r_s}{r_w}, \tag{6-1}$$

$$r_{wa} = r_w e^{-s}. \tag{6-2}$$

The main fracture parameter—fracture length (L) is defined as $L = 2r_{wa}$.

A well with a negative value of the skin requires much smaller pressure drop in comparison with a well with a positive skin factor to maintain a constant fluid flow rate (please see the columns 5 and 6 in Table 6-3). For hydraulically fractured wells the r_{wa}/r_w ratio can be very large. We will assume that for stimulated wells ($s < 0$) the apparent radius concept can be used. In practice this means that the drainage radius (Eppelbaum and Kutasov 2006a) should be considerably larger than the apparent radius. This chapter describes a new method for determining the skin factor that enables to evaluate the efficiency of wellbore stimulation operations (hydraulic fracturing or acidacion). For this reason we assume that the formation permeability was estimated from previous buildup or drawdown pressure well tests.

6.1 A Well as a Cylindrical Source

For stimulated wells the value of dimensionless time based on the apparent radius (Eq. (6-2)) can be large and the wellbore should be considered to be a cylindrical source.

Let's assume that a well is producing at a constant flow rate from an infinite-acting reservoir with the following characteristics: the reservoir is a homogeneous and isotropic porous medium of uniform thickness, the porosity and permeability are constant, the fluid is small and has constant compressibility, constant fluid viscosity, small pressure gradients, negligible gravity forces and laminar flow in the reservoir and the Darcy law can be applied. For a well as a cylindrical source we need to obtain the solution to the diffusivity equation under the following boundary and initial conditions:

$$p(t=0,r) = p_i \; r_w \le r < \infty \quad t > 0 : \quad \left(r \frac{\partial p}{\partial r} \right)_{r_w} = \frac{q\mu}{2\pi k h}, \qquad t > 0 . \tag{6-3}$$

$$p(t,r \to \infty) \to p_i \qquad t > 0 . \tag{6-4}$$

where h is the reservoir thickness, q is the fluid flow rate, p_i is the initial reservoir pressure, μ is the viscosity and t is the production/test time.

It should be noted that conventionally in reservoir engineering the produced flow rate q (Eq. (6-3)) has a positive sign and negative sign is applied to injection wells. It is well known that in this case, the solution to the diffusivity equation is a complex integral (Van Everdingen and Hurst 1949; Carslaw and Jaeger 1959). For the bottom-hole pressure (in terms of pressure and fluid flow rate) we obtained the following semi-analytical equation (Kutasov 2003)

$$p_{wf}(t) = p(t,r_w) = p_i - \frac{q\mu}{2\pi k} \ln\left[1 + \left(c - \frac{1}{a + \sqrt{t_{Da}}} \right) \sqrt{t_{Da}} \right], \tag{6-5}$$

$$t_{Da} = \frac{0.0002637 kt}{\phi c_t \mu r_{wa}^2}, \quad a = 2.7010505, \quad c = 1.4986055, \tag{6-6}$$

where ϕ is the porosity, c_t is the total compressibility and p_{wf} is the bottom-hole flowing pressure, psi.

Let's introduce the dimensionless bottom-hole pressure

$$p_D(t_{Da}) = \frac{2\pi k h (p_i - p_w)}{q_h \mu}. \tag{6-7}$$

Then

$$p_D\left(t_{Da}\right) = \ln\left[1 + \left(c - \frac{1}{a + \sqrt{t_{Da}}}\right)\sqrt{t_{Da}}\right].$$
(6-8)

Values of p_D calculated from Eq. (6-8) and the results of a numerical solution ("Exact" solution) by Chatas (Lee 1982, p.107) were compared (Kutasov 2003). The agreement between values of p_D calculated by these two methods was very good. For this reason the principle of superposition can be used without any limitations.

Eq. (6-8) can be rewritten as

$$\exp\left[p_D\left(t_{Da}\right)\right] - 1 = \left(c - \frac{1}{a + \sqrt{t_{Da}}}\right)\sqrt{t_{Da}}.$$
(6-9)

Let

$$b = \exp\left[p_D\left(t_{Da}\right)\right], \qquad x = \sqrt{t_{Da}}.$$
(6-10)

Then

$$x^2 + a_1 x + a_2 = 0.$$
(6-11)

The solution of Eq. (6-11) is

$$x = \sqrt{t_{Da}} = -\frac{a_1}{2} + \sqrt{\frac{a_1^2}{4} - a_2}, \qquad a_1 = \frac{ca - b - 1}{c}, \qquad a_2 = -\frac{ba}{c}.$$
(6-12)

The apparent (effective) well radius skin factor is calculated from Eqs. (6-2) and (6-6)

$$r_{wa} = \sqrt{\frac{0.0002637kt}{t_{Da}\phi c_t \mu}}, \qquad s = -\ln\frac{r_{wa}}{r_w}.$$
(6-13)

This yields a quadratic equation for estimating the skin factor. It is interesting to note that we obtained a similar equation for the calculation of the skin factor for a well produced at a constant bottom-hole pressure (Kutasov and Kagan 2003b).

6.2 Example

It is routinely accepted that the first step in validation of a suggested new method is to use an "Exact" (numerical) solution and generate data for a simulated test. Subsequently a semi-analytical equation is used to process the results of the simulated test and compare the obtained and assumed input parameters. Below we used the results of a field drawdown well test (Earlougher 1977, Example 3.1). The reservoir and test data are presented in Table 6-1. It was found that the deviation from Horner plot caused by the skin and borehole storage effects was observed during the first 4 hours. The duration of the transient period was 12 hours and at $t > 16$ hours the pressure behavior was influenced by reservoir boundaries, interference from nearby wells, neighboring boundaries, interference from nearby wells, and/or other factors. Let us now assume that after well stimulation the skin factor was lowered from $s = 4.6$ to $s = -5.108$.

Using the Table C.1 (Lee 1982, p. 107) of $p_D = f(t_{Da})$ we generated data for this simulated example (Table 6-2).

Table 6-1. Reservoir data. B is the oil formation volume factor.

$h = 130$ ft	$\phi = 0.20$
$r_w = 0.25$ ft	$\mu = 3.93$ cP
$q = 348$ STB/D	$p_i = 1,154$ psi
$B = 1.14$ RB/STB	$k = 89$ md
$c_t = 8.74 \cdot 10^{-6}$ 1/psi	$S = 4.6$

Table 6-2. Calculated values of the skin factor and pressure drops for a stimulated well: t_{Da} ($t = 1$ hr) $= 2.0$; $s = -5.108$ (assumed).

t, hrs	p_D	t_{Da}	s	Δp, psi; $s = -5.108$	Δp, psi; $s = 4.60$
1.0	1.0195	1.992	−5.110	19.40	199.01
1.5	1.1665	3.015	−5.105	22.20	202.86
2.0	1.2750	4.020	−5.105	24.26	205.60
2.5	1.3625	5.025	−5.105	25.92	207.72
3.0	1.4362	6.033	−5.105	27.33	209.46
3.5	1.4997	7.039	−5.105	28.54	210.93
4.0	1.5557	8.045	−5.105	29.60	212.20
4.5	1.6057	9.050	−5.105	30.55	213.32
5.0	1.6509	10.053	−5.105	31.41	214.32
7.5	1.8294	15.082	−5.105	34.81	218.18
10.0	1.9601	20.133	−5.105	37.30	220.91
15.0	2.1470	30.155	−5.105	40.85	224.77

The value of $s = -5.108$ was selected to avoid interpolation of p_D values. This yielded:

$$t_{Da}(t = 1 \text{ hr}) = \frac{0.0002637 \cdot 89 \cdot 1}{0.20 \cdot 8.74 \cdot 10^{-6} \cdot 3.93 \cdot 0.25^2 \exp(2 \cdot 5.108)} = 2$$

The results of Eqs. (6-6), (6-12) and (6-13) application are presented in Table 6-2. The calculations show that Eq. (6-5) can be used to compute the flow pressure and skin factor. The assumed and calculated values of the skin factor are in very good agreement.

We can conclude that a new method for analyzing the constant flow rate pressure test data was presented. This method can be used to calculate the skin factor for stimulated wells. It can be applied by oilfield service companies to evaluate the efficiency of wellbore stimulation operations (hydraulic fracturing or acidacion).

7

Designing an Interference Well Test in a Geothermal Reservoir

7.1 Determination of the Formation Permeability, Hydraulic Diffusivity and the Porosity

A new technique has been developed for determination of the formation permeability, hydraulic diffusivity, and the porosity—total compressibility product from interference well tests in geothermal reservoirs (Kutasov and Eppelbaum 2008). At present the curve-matching technique (where only values of several pressure drops are used for matching) is utilized to process field data. A new method of field data processing, where all measured pressure drops are utilized, is proposed. It is also shown that when high precision (resolution) pressure gauges are employed the pressure time derivative equation can be used for determination of formation hydraulic diffusivity. An example of designing an interference test is presented. A field example is also presented to demonstrate the data processing procedure.

The standard drawdown and buildup pressure well tests are used to determine the formation permeability and to estimate to what degree (expressed through skin factor) the drilling and production operations altered the permeability of formations near the wellbore. To process the field data the formation porosity (φ)—total compressibility (c_t) product should be known. These parameters cannot be determined from a pressure or flow test in a singular well. Only from interference tests (multiple-well tests) the value of φc_t can be estimated. The name comes from the fact that the pressure drop caused by the producing wells at the closed observational well "interferes with" the pressure at the observational well (Matthews and Russell 1967). In interference testing at least two wells are used: one well (the "active" well) is put on production or injection, and in the second "observational" well the pressure changes are observed during production (injection) and shut-in of the first well. The pressure response in the "observational" well enables estimation of the thickness (h)—formation permeability (k) product (reservoir transmissivity) and the hydraulic diffusivity of formations (η). After the value of η is determined the formation porosity—total compressibility product can be estimated. To process interference test data the type-curve matching technique is used determined by the relative error of the ratio ψ (Eqs. (7-9) and (7-10)). For example, if the value of the relative error $\Delta\psi/\psi$ is 0.001, then $\varepsilon = 0.001$. Note that if N records of pressure drops are available, it is possible to obtain $N \cdot (N-1)/2$ values of D. In this case the regression technique can be used to analyze test data.

From Eqs. (7-7) or (7-8) we can calculate the parameter M and, hence, the value of formation permeability

$$k = 70.6\frac{q\mu B}{Mh}. \tag{7-11}$$

The coefficient of hydraulic diffusivity and the total compressibility-porosity product can be determined from Eqs. (7-2) and (7-3)

$$\eta = \frac{R^2}{4D}, \qquad \phi c_t = \frac{0.0002637k}{\eta\mu}. \tag{7-12}$$

7.3 Drainage Radius

Presented solutions of the diffusivity differential equation are valid only for the transient (infinite-acting reservoir) period. This means that during the test the pressure field around the borehole is practically not affected by the reservoir's boundaries or by other production (injection) wells. Therefore, in order to estimate the duration of the transient period, the drainage radius should be determined with a sufficient accuracy. It is desirable to have an approximate relationship between dimensionless cumulative production and dimensionless time (e.g., Kutasov and Hejri 1984; Johnson 1986). We used the material balance condition to determine the well drainage radius. For a well producing at a constant flow rate (Eppelbaum and Kutasov 2006a) it was found that the following equation could be used to estimate the dimensionless drainage radius as a function of dimensionless production or injection time

$$t_D = \frac{1}{4}\frac{R_{dr}^2 - 2\ln\left(R_{dr}\right) - 1}{\ln\left(R_{dr}\right)} \cdot \ln\left[1 + \left(c - \frac{1}{a + \sqrt{t_D}}\right)\sqrt{t_D}\right], \tag{7-13}$$

where

$$\begin{cases} t_D = \dfrac{kt}{\phi c_t \mu r_w^2}, \quad R_{dr} = \dfrac{r_{dr}}{r_w}, \\ a = 2.7010505; c = 1.4986055 \end{cases} \tag{7-14}$$

It should be noted that here dimensionless time is based on the active well radius and the values of t_D are very large. Calculations after Eq. (7-13) show that the function $R_{dr} = f(t_D)$ can be approximated by a simple formula:

$$R_{dr} = 1 + d\sqrt{t_D}. \tag{7-15}$$

For $10^3 < t_D < 10^{16}$ the values of d vary from 2.053 to 2.017 (for $t_D = 10^{16}$), and therefore, for practical purposes, we can assume that:

$$R_{dr} \approx 2\sqrt{t_D} \qquad \text{or} \qquad r_{dr} \approx 2\sqrt{\eta t}. \tag{7-16}$$

7.4 Test Designing Example

When designing interference well test preference should be given to a producing "active" well rather than to an injection well. The application of the above mentioned working equation requires that the physical properties of the injected fluids (dynamic viscosity, compressibility, volumetric thermal expansion) are the same as the properties of the reservoir fluids. For example, viscosity of water is very much dependent on the temperature. Thus the mobility (formation permeability–viscosity ratio) of the injected water should be very close to that of the reservoir water. The transient test analysis for an injection well with non-unit mobility ratio is complex and the reservoir should be considered as a composite system (Earlougher 1977). Let us assume that during an interference test, water with temperature of 80°C (176°F) was produced in Well 1 for 50 hours. The pressure response in Well 2, 150 ft away, was observed for 100 hours. Known and estimated reservoir properties are presented in Table 7-1.

Table 7-1. Reservoir data.

$h = 20$ ft	depth $= 4{,}920$ ft
$t_p = 50$ hours	$\mu = 0.355$ cP
$q = 400$ STB/D	$R = 150$ ft
$B = 1.014$ RB/STB	$p_i = 649.74$ psia
$c_t = (3.0\text{–}4.5)\cdot 10^{-5}$ 1/psia	$\varphi = 0.2\text{–}0.3$
$T = 176°F$	$k = 100\text{–}150$ mD

We assumed that the values φ, c_t, and k might vary in some intervals (Table 7-1). The value of dynamic viscosity was taken from Table 7-2.

Table 7-2. Dynamic viscosity of water (Internet: http://www.engineeringtoolbox.com).

T,°C	μ, cP	T, °C	μ, cP	T, °C	μ, cP
0	1.78	35	0.719	70	0.404
5	1.52	40	0.653	75	0.378
10	1.31	45	0.596	80	0.355
15	1.14	50	0.547	85	0.334
20	1.00	55	0.504	90	0.314
25	0.890	60	0.467	95	0.297
30	0.798	65	0.434	100	0.281

The water formation volume factor (B) was calculated from the following equation (Kutasov 1989b):

$$B = 0.998 \exp\left[-\alpha p - \beta(T - T_s) - \gamma(T - T_s)^2\right], \qquad (7\text{-}17)$$

where pressure is in psig, temperature in °F, $T_s = 59°F$ and

$\alpha = 2.7384 \cdot 10^{-6}\, 1/\text{psig},$

$\beta = -1.5353 \cdot 10^{-4} \,°F^{-1}, \gamma = -7.4690 \cdot 10^{-7} \,°F^{-2}.$

The maximum and minimum values of the hydraulic diffusivity coefficient are:

$$\eta = \frac{0.0002637 \cdot 150}{0.20 \cdot 3 \cdot 10^{-5} \cdot 0.355} = 18,570 \left(\frac{ft^2}{hr}\right),$$

$$\eta = \frac{0.0002637 \cdot 100}{0.30 \cdot 4.5 \cdot 10^{-5} \cdot 0.355} = 5,502 \left(\frac{ft^2}{hr}\right).$$

The corresponding values of the parameter D are:

$$D = \frac{150^2}{4 \cdot 18,570} = 0.3029(\text{hrs}),$$

$$D = \frac{150^2}{4 \cdot 5,502} = 1.0223(\text{hrs}).$$

The pressure drops in the observational well and values of Δt_x were calculated after Eqs. (7-1), (7-4) and (7-6) are presented in Table 7-3.

Table 7-3. The pressure drops and values of Δt_x.

t, hours	$D = 1.0223$, $\Delta t_x = 0.1811$, hours	$D = 0.3029$, $\Delta t_x = 0.0429$, hours
	Δp, psia	Δp, psia
3.0	4.13	6.15
5.0	6.12	7.75
10.0	9.17	10.00
20.0	12.44	12.30
30.0	14.42	13.65
40.0	15.84	14.62
49.9	16.94	15.36
50.1	16.96	15.34
51.0	15.97	12.39
60.0	8.69	5.99
70.0	6.19	4.21
80.0	4.88	3.30
90.0	4.05	2.73
100.0	3.47	2.34

Thus the predicted values of the pressure drops will be within the 0–17.0 psia interval. It will be shown in the next section that to determine with high accuracy (by monitoring the pressure drops in the observational well) the values of Δt_x, pressure gauges with high limiting precision (resolution) are needed.

7.5 Field Case

During an interference test, water was injected into Well A for 48 hours. The pressure response in Well B, 119 ft away, was observed for 148 hours. Known reservoir properties are presented in Table 7-4 (Earlougher 1977, Example 9.1).

Table 7-4. Reservoir data, field case.

$h = 45$ ft	depth = 2,000 ft
$t_p = 48$ hours	$\mu = 1.0$ cP
$q = -170$ STB/D	$R = 119$ ft
$B = 1.0$ RB/STB	$p_i = 0$ psig
$c_t = 9.0 \cdot 10^{-6}$ 1/psia	

The observed pressure and the flow history are presented in Table 7-5 and Figure 7-1.

Table 7-5. Interference test data for the field example.

t, hours	p_w, psig	$\Delta p = p_i - p_w$, psia
0.0	0	-
4.3	22	-22
21.6	82	-82
28.2	95	-95
45.0	119	-119
48.0	Injection ends	
51.0	109	-109
69.0	55	-55
73.0	47	-47
93.0	32	-32
142.0	16	-16
148.0	15	-15

Earlougher (1977) used the curve-matching method and also analyzed the data by three different computer-aided techniques. It was found that it would probably be best to use $k = 5.7$ mD and $\varphi c_t = 0.99 \cdot 10^{-6}$ 1/psia. The parameter D is:

$$D = D_E = \frac{119^2 \cdot 0.99 \cdot 10^{-6} \cdot 1.0}{4 \cdot 0.0002637 \cdot 5.7} = 2.331 \, (\text{hrs}).$$

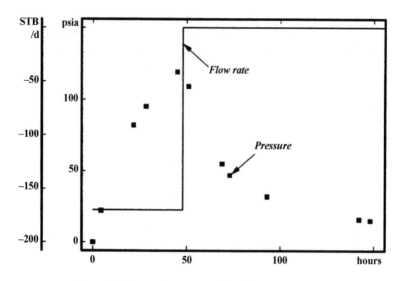

Figure 7-1. Observed pressure and flow history (Horne 1995).

Horne (1995) analyzed the same field data and for the matching point at $t = 24$ hours the values of $k = 5.507$ mD, $\varphi = 0.109$ were determined, and

$$D = D_H = \frac{119^2 \cdot 0.109 \cdot 9 \cdot 10^{-6} \cdot 1.0}{4 \cdot 0.0002637 \cdot 5.507} = 2.392 (\text{hrs}).$$

We used time pairs 4.3–21.6 hrs, 4.3–28.2 hrs and 4.3–45.0 hrs (Table 7-5) and from Eqs. (7-7), (7-10) and (7-12) computed the squared averaged values: $k = 5.222$ mD, and $\varphi = 0.109$.

Then

$$D = D_{KE} = \frac{119^2 \cdot 0.117 \cdot 9 \cdot 10^{-6} \cdot 1.0}{4 \cdot 0.0002637 \cdot 5.222} = 2.707 (\text{hrs}).$$

Now the shut-in time at which the maximum pressure drop occurs can be determined from Eq. (7-6) (Figure 7-2).

The transient pressure drops were calculated (Eqs. (7-1) and (7-4)) for three sets of parameters (k, φc_t) obtained by Earlougher (1977); Horne (1995) and by the authors. The pressure drops are denoted as Δp_E, Δp_H and Δp, correspondingly (Table 7-6).

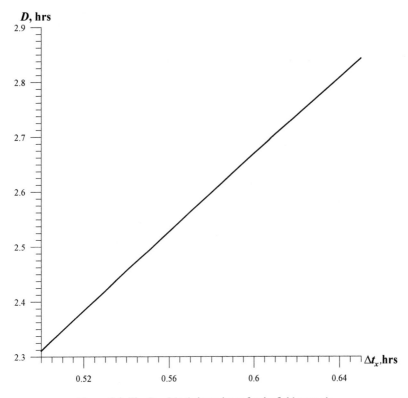

Figure 7-2. The $D = D(\Delta t_x)$ dependence for the field example.

Table 7-6. Comparison of observed pressure changes, Δp^*, with calculated pressure drops (Δp_E, Δp_H, Δp).

t, hours	Δp^*, psia	Δp_E, Psia	Δp_H, psia	Δp, psia
4.3	22	23.9	24.0	22.0
21.6	82	82.1	83.7	83.0
28.2	95	93.4	95.4	95.2
45.0	119	113.9	116.6	117.3
51.0	109	104.3	107.3	110.0
69.0	55	52.2	53.9	56.4
73.0	47	47.4	49.0	51.2
93.0	32	32.7	33.9	35.5
142.0	16	18.9	19.6	20.6
148.0	15	18.0	18.6	19.6

From Table 7-6 one can see that the calculated pressure drops are in satisfactory agreement (taking also into account the low accuracy of Δp^* values) with the observed ones. For the parameter of $D = D_E = 2.3310$ hrs the computed (from Eq. (7-6)) value of Δt_x is 0.5054 hrs or 30.32 min (Figure 7-2). The maximum pressure drop (after Eq. (7-4)) is 117.191 psia (Table 7-6). Now let us assume that the value of Δt_x can be detected with accuracy of ± 0.5 minutes and the calculated values of D (after Eq. (7-6)) are presented in Table 7-7.

Table 7-7. The accuracy of determination of the parameter D, $\Delta t_x = 30.32$ min.

$\Delta t - \Delta t_x$, minutes	Δp, psia	D, hours	$D/2.3310$
−5	117.170	2.0193	0.866
−4	117.177	2.0827	0.894
−3	117.183	2.1456	0.921
−2	117.187	2.2079	0.947
−1	117.190	2.2697	0.974
0	117.191	2.3310	1.000
+1	117.190	2.3917	1.026
+2	117.187	2.4520	1.052
+3	117.182	2.5118	1.078
+4	117.176	2.5711	1.103
+5	117.167	2.6301	1.128

In conclusion we state that at present the type-curve matching technique is used to estimate the porosity-total compressibility product (φc_t) and formation permeability (k) from interference well tests. The disadvantage of this method is that (usually) only several values of pressure drops can be used. A more effective method for processing results of well interference tests is suggested. The basic equation (the exponential integral) remains the same. However, *all pressure drops points* can be used to obtain averaged values of φc_t and k. It is also shown that in some cases (when high resolution electronic gauges are used) the time derivative of the transient pressure can be utilized.

In this case pressure gauges should have resolution of ± 0.02 psia, and even in this case the hydraulic diffusivity can be determined with the accuracy of $\pm 13\%$. It is clear that in this field example Eq. (7-6) cannot be used because the accuracy of measurements is about ± 0.5 psia (Table 7-5).

8

Interference Well Testing–Variable Fluid Flow Rate

At present for conducting an interference well test a constant flow rate (at the "active" well) is utilized and the type-curve matching technique (where only values of several pressure drops are matched) is used to estimate the porosity-total compressibility product and formation permeability. For oil and geothermal reservoirs with low formation permeability the duration of the test may require a long period of time and it can be difficult to maintain a constant flow rate. In this study we present working equations which will allow processing field data when the flow rate at the "active" well is a function of time (Kutasov et al. 2008). The shut-in period is also considered. As was mentioned in Chapter 7 the drawdown and buildup pressure well tests are used to determine the formation permeability and to estimate to what degree (expressed through skin factor) the drilling and production operations altered the permeability of formations near the wellbore. To process the field data the formation porosity—total compressibility product should be known. These parameters cannot be determined from a pressure or flow test in a singular well. Only from interference tests (multiple-well tests) the necessary parameters can be estimated.

The pressure response in the "observational" well, allows estimation of the thickness (h)—formation permeability (k) product (reservoir transmissivity) and the hydraulic diffusivity of formations (μ). After the value of η is determined the formation porosity—total compressibility product can be estimated. To process interference test data the type-curve matching technique is used (Earlougher 1977; Horne 1995). This technique is based on the assumption that during production (or injection) the radial flow behavior in an infinite gomogeneous reservoir can be expressed by the Ei function —the exponential integral. The set of data obtained after shut-in of the "active" well can be matched to a special form of the exponential integral or line source type curve developed by H.J. Ramey (Jr.) (Horne 1995). The disadvantage of the type-curve matching technique is that, usually, only several values of pressure drops can be used. It should be also noted that pressure drops can be very small over distance and high accuracy electronic gauges should be utilized to monitor the pressure changes in the "observational" well.

We should also make a general comment: while designing interference well test the preference should be given to a producing "active" well rather than to an injection well. The application of the above mentioned working equation requires that the physical properties of the injected fluids (dynamic viscosity, compressibility, volumetric thermal expansion) are the same as the properties of the reservoir fluids. For example, viscosity of water is very much dependent on the temperature. Thus the mobility (formation permeability–viscosity ratio) of the injected water should be very close to that of the reservoir water. The transient test analysis for an injection well with non-unit mobility ratio is complex and the reservoir should be considered as a composite system (Earlougher 1977).

For oil and geothermal reservoirs with low formation permeability the duration of the test may require a long period of time and it can be difficult to maintain a constant flow rate. In this study we present working equations which will allow to process field data when the flow rate at the "active" well is a function of time. The shut-in period is also considered.

Thus the objective of this analysis is twofold: to propose a new techniques for field data processing; and present working formulas for interference tests conducted at a variable fluid flow rate at the "active" well.

8.1 Constant Flow Rate—the Basic Equation

For $r = R$ and $t < t_1$ (Matthews and Russell 1967)

$$p_i - p_w(t) = -M\,Ei\left(-\frac{D}{t}\right). \tag{8-1}$$

Introducing the dimensionless pressure drop during the fluid production (injection) period

$$p_{wD} = \frac{p_i - p_w(t)}{M} = -Ei\left(-\frac{D}{t}\right), \tag{8-2}$$

$$D = \frac{R^2}{4\eta},\quad M = 70.6\frac{q_o\mu B}{kh},\quad \eta = \frac{0.0002637k}{\phi c_t \mu}, \tag{8-3}$$

where p_i is the initial pressure and p_w is the bottom-hole pressure at the "observational" well.

For $r = R$ and $t > t_1$

$$p_{ws}(R,t) = p_i - M\left[-Ei\left(-\frac{D}{t}\right) + Ei\left(-\frac{D}{t-t_1}\right)\right], \tag{8-4}$$

where p_{ws} is the bottom-hole pressure at the "observational" well after shut-in of the "active" well.

Below we will also show that the time derivative of p_{ws} (in some cases) can be utilized at designing an interference test. The differentiation of the last equation yields

$$p_{ws}^{\bullet}(t) = \frac{dp_{ws}}{dt} = -\frac{M}{t-t_1}\exp\left(-\frac{D}{t-t_1}\right) + \frac{M}{t}\exp\left(-\frac{D}{t}\right). \qquad (8\text{-}5)$$

Let us now assume that at some time $t = t_x$ the pressure drop at the observational well reaches its maximum value. Then $p_{ws}^{*} = 0$, and from the last equation we obtain

$$\ln\frac{t_x}{t_x-t_1} = D\left(\frac{1}{t_x-t_1}-\frac{1}{t_x}\right), \qquad \Delta t_x = t_x - t_1. \qquad (8\text{-}6)$$

Thus, by specifying the parameter D, we can evaluate the time when the maximum pressure drop occurs at the observational well.

8.2 Variable Flow Rate

We have chosen the quadratic function (Figure 8-1) to approximate the production (or injection) fluid flow rate (F) at the "active" well. This function allows one to design interference tests when the flow rate reaches its maximum or minimum value. In this case the corresponding pressure drops and time delays (at the "observation" well) can be utilized for determination of formation parameters (k, φc, μ). Let as assume that during production (injection) period

$$F(t) = q_0 q, \qquad q(t) = a_0 + a_1 t + a_2 t^2, \qquad 0 < t \le t_1,$$

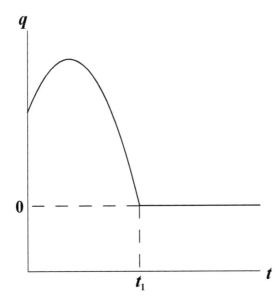

Figure 8-1. Fluid flow rate versus time—a schematic curve.

where a_0, a_1, a_2 are some coefficients. For shut-in period

$$q(t) = 0, \qquad t > t_1.$$

It is well known that the superposition theorem (Duhamel integral) can be used to derive solutions for time-dependent boundary conditions (Carslaw and Jaeger 1959). In our case for the interval $0 \le t \le t_1$ have to find the following integral:

$$y_1(t) = \frac{p_v(t) - p_i}{M} = \int_0^t q(\tau) \frac{d}{dt} p_{wD}(t - \tau) d\tau, \tag{8-7}$$

$$p_{wD}(t - \tau) = -Ei\left(-\frac{D}{t - \tau}\right), \tag{8-8}$$

where p_{wD} is the dimensionless pressure in the observational well for the constant-rate case (Eq. (8-2)), τ is the variable of integration, and D is a positive constant.

The derivative of this function with respect to time has the form:

$$\frac{d}{dt} p_{wD}(t - \tau) = -\frac{\exp\left(-\frac{D}{t - \tau}\right)}{t - \tau}. \tag{8-9}$$

Our calculations provide us with the following result. In case when $0 \le t \le t_1$ we have

$$y_1(t) = \frac{p_w(t) - p_i}{M} = -\frac{1}{2}\left(2a_1 t + a_2 tD + 3a_2 t^2\right)\exp\left(-\frac{D}{t}\right) - \\ \left(a_0 + a_1 D + \frac{1}{2} a_2 D^2 + a_2 t^2 + 2a_2 Dt + a_1 t\right)Ei\left(-\frac{D}{t}\right) \tag{8-10}$$

The corresponding time derivative for this case:

$$u_1(t) = \frac{dy_1}{dt} = -\left[2a_2(t + D) + a_1\right]Ei\left(-\frac{D}{t}\right) - \exp\left(-\frac{D}{t}\right)\left(2a_2 t - \frac{a_0}{t}\right). \tag{8-11}$$

On the next step we consider the case when $q(t) = 0$ for $t \ge t_1$. Here we also calculate the integral

$$y_2(t) = \frac{p_{ws}(t) - p_i}{M} = \int_0^t q(\tau) \frac{d}{dt} p_{wD}(t - \tau) d\tau - \int_{t_1}^t q(\tau) \frac{d}{dt} p_{wD}(t - \tau) d\tau \\ = \int_0^{t_1} q(\tau) \frac{d}{dt} p_{wD}(t - \tau) d\tau \tag{8-12}$$

In the case $t \geq t_1$, we have

$$
y_2(t) = \left[a_0 + a_1(t+D) + a_2(t+D)^2 - \frac{D^2}{2} \right] \left(Ei\left(-\frac{D}{t-t_1}\right) - Ei\left(-\frac{D}{t}\right) \right)
$$
$$
+ \left\{ a_1(t-t_1) + \frac{1}{2} a_2(3t-t_1)^2 + (4t_1+D)(t-t_1) \right\} \exp\left(-\frac{D}{t-t_1}\right) \tag{8-13}
$$
$$
- \left(a_1 t + \frac{1}{2} a_2(3t^2+Dt) \right) \exp\left(-\frac{D}{t}\right)
$$

A commercially available software, Maple 7 (Waterloo Maple 2001), was utilized to compute functions $y_1(t)$, $y_2(t)$ and $u_1(t)$.

8.3 Working Equations

At designing interference well test the range of probable values of the parameter D can be estimated. After this the time ($t = t_x$) when the maximum (or minimum) pressure drops occur at the observational well can be found from equations:

$$
y_1(t = t_x) = 0, y_2(t = t_x) = 0 \text{ or } u_1(t = t_x) = 0.
$$

At least two measurements of transient pressure at the observational well (at time $t = t_a$ and $t = t_b$) are needed to calculate the coefficient of hydraulic diffusivity, formation permeability, and the total compressibility-porosity product.

Let (see Eq. (8-10))

$$
y_1(t = t_a) = \frac{P_w(t_a) - P_i}{M} = g(D, t_a), \tag{8-14}
$$

$$
y_1(t = t_b) = \frac{P_w(t_b) - P_i}{M} = g(D, t_b). \tag{8-15}
$$

Then the pressure drop ratio is

$$
\psi = \frac{P_i - P_w(t_a)}{P_i - P_w(t_b)} = \frac{g(D, t_a)}{g(D, t_b)} = f(D). \tag{8-16}
$$

Although Eq. (8-16) is based on an analytical solution (Eq. (8-10)) we should note several limitations in the suggested method application. Firstly, the pressure drop ratio ψ (Eq. 8-16) should be determined with high accuracy. This means that high accuracy of pressure drops measurements is needed. The pressure differences $P_i - P_w(t_a)$ and $P_i - P_w(t_b)$ should be significantly larger than the absolute accuracy of pressure measurements. Let's assume that the absolute accuracy of the ratio ψ is ε and then solving the following equation we calculate the value of D

$$\psi - f(D) = \varepsilon. \tag{8-17}$$

The Newton method was used for solving Eq. (8-10) (Grossman 1977). In this method a solution of an equation is sought by defining a sequence of numbers which become successively closer and closer to the solution. The conditions, which guarantee that Newton method in our case will work and provide a unique solution, are satisfied (Grossman 1977, p. 259). The selection of the parameter ε (Eq. (8-17)) is determined by the relative error of the ratio ψ (Eqs. (8-16) and (8-10)). For example, if the value of the relative error $\Delta\psi/\psi$ is 0.001, then $\varepsilon = 0.001$. Note that if N records of pressure drops are available, it is possible to obtain $N \cdot (N-1)/2$ values of D. In this case the regression technique can be used to analyze test data. From Eqs. (8-14) or (8-15) we can calculate the parameter M and, hence, the value of formation permeability

$$k = 70.6 \frac{q\mu B}{Mh}. \tag{8-18}$$

The coefficient of hydraulic diffusivity and the total compressibility-porosity product can be determined from Eq. (8-3):

$$\eta = \frac{R^2}{4D}, \qquad \phi c_t = \frac{0.0002637k}{\eta\mu}. \tag{8-19}$$

Example

Problem: An interference test was run in shallow-water sand (Lee 1982, pp. 90–91). The active well, Well 13, produced 466 STB/D water. Pressure response in shut-in Well 14, which was 99 ft from Well 13, was measured as a function of time elapsed since the drawdown in Well 13 began. Estimated rock and fluid properties include $\mu = 1.0$ cP, $B = 1.0$ RB/STB, $h = 9$ ft, $r_w = 3$ in, and $\varphi = 0.3$. Total compressibility is unknown. Pressure readings in Well 14 were given in Table 8-1. Estimate formation permeability and total compressibility. The observed pressure and the reservoir data presented in Tables 8-1 and 8-2. Lee (Lee 1982) used the curve-matching method. It was found that (for the match point at $t = 128$ minutes) $k = 1,433$ mD and $c_t = 2.74 \cdot 10^{-5}$ 1/psia.

For a constant flow rate ($a_0 = 1$, $a_1 = a_2 = 0$) Eq. (8-16) transforms to

$$\psi = \frac{p_i - p_w(t_a)}{p_i - p_w(t_b)} = \frac{Ei\left(-\dfrac{D}{t_a}\right)}{Ei\left(-\dfrac{D}{t_b}\right)} = f(D). \tag{8-20}$$

From the last equation the parameter D was computed and from Eq. (8-1) (at $t = t_a$ or $t = t_b$) the ratio M can be determined. After this values of k, η and φc_t are estimated from Eqs. (8-18), (8-19) and (Table 8-3). The averaged squared values of the computed (Table 8-3) parameters are:

$$k = 1,413 \pm 38 \text{ mD}, \qquad \eta = 45,167 \pm 3,671 \text{ ft}^2/\text{hr},$$

$$\varphi c_t = (0.8319 \pm 0.0449) \cdot 10^{-5} \text{ 1/psia}.$$

The total compressibility (at $\varphi = 0.3$) is

$$c_t = (2.273 \pm 0.150) \cdot 10^{-5} \text{ 1/psia}.$$

The parameter for this example is

$$D = \frac{99^2}{4 \cdot 45,167} = 0.0543 \text{ (hrs)}.$$

We calculated the values of pressure drops during production and shut-in period (after Eqs. (8-1) and (8-4)). From Figure 8-2 one can see that the calculated pressure drops are in good agreement with observed ones (Table 8-1). For the parameter of $D = 0.0543$ hrs the computed (from Eq. (8-6)) value of Δt_x is 0.00758 hrs or 0.455 min. The maximum pressure drop (after Eq. 8-4) is 11.9346 psia (Table 8-4). Now let assume that the value of Δt_x can be detected with accuracy of ± 0.15 minutes and the calculated values of D (after Eq. (8-6)) are presented in Table 8-4. In this case pressure gauges should have resolution of ± 0.0003 psia, and even in this case the hydraulic diffusivity can be determined with the accuracy of $\pm 20\%$. It is clear that in this example Eq. (8-6) cannot be used for estimation of the hydraulic diffusivity because the accuracy of measurements is about ± 0.01 psia (Table 8-1).

Table 8-1. Interference test data for the example.

T, min	p_w, psig	$\Delta p = p_i - p_w$, psia
0.0	148.92	0
25	144.91	4.01
40	143.72	5.20
50	143.18	5.74
100	141.47	7.45
200	139.72	9.20
300	138.70	10.22
400	137.99	10.93
580	137.12	11.80

Table 8-2. Reservoir data.

$h = 9$ ft	$\varphi = 0.3$
$t_i = 580$ min	$\mu = 1.0$ cP
$q_o = 466$ STB/D	$R = 99$ ft
$B = 1.0$ RB/STB	$p_i = 148.92$ psig
$r_w = 3$ in	

Table 8-3. Computed values of formation permeability (k), hydraulic diffusivity (η), and the porosity-total compressibility product (φc_t).

t_a, min	t_b, min	Δp_a, psia	Δp_b, psia	k, mD	η, ft²/hr	φc_t, 10^{-5} 1/psia
25	50	4.01	5.74	1310	37959	0.9099
25	100	4.01	7.45	1363	40634	0.8847
25	200	4.01	9.20	1382	41600	0.8759
25	300	4.01	10.22	1391	42064	0.8717
25	400	4.01	10.93	1399	42502	0.8678
25	580	4.01	11.80	1418	43521	0.8589
50	100	5.74	7.45	1415	45291	0.8236
50	200	5.74	9.20	1415	45291	0.8236
50	300	5.74	10.22	1419	45628	0.8201
50	400	5.74	10.93	1426	46188	0.8143
50	580	5.74	11.80	1445	47624	0.7999
100	200	7.45	9.20	1414	45291	0.8236
100	300	7.45	10.22	1422	46014	0.8149
100	400	7.45	10.93	1432	47030	0.8031
100	580	7.45	11.80	1457	49500	0.7760
200	580	9.20	11.80	1484	54149	0.7229

Table 8-4. The pressure drops versus shut-in time.

Δt, min	D, hrs	$D/0.0543$	$p_i - p_{ws}$, psia
0.300	0.0379	0.698	11.9342
0.350	0.0433	0.797	11.9344
0.400	0.0486	0.895	11.9346
0.455	0.0543	1.000	11.9346
0.500	0.0589	1.085	11.9346
0.550	0.0639	1.177	11.9343
0.600	0.0688	1.267	11.9337

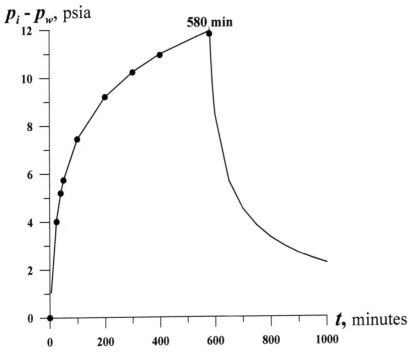

Figure 8-2. Pressure drop versus time.

Some conclusions

For reservoirs with low formation permeability the duration of the test may require a long period of time and it can be difficult to maintain a constant flow rate. We developed working equations which will enable the processing of field data when the flow rate at the "active" well is a quadratic function of time. At present the type-curve matching technique is used to estimate the porosity-total compressibility product and formation permeability from interference well tests. The disadvantage of this method is that only several values of pressure drops can be used. A more effective method for processing results of well interference tests is suggested where *all pressure drops points* are used to obtain averaged values of φc_t and k.

9

Determination of Formation Permeability and Skin Factor from Afterflow Pressure and Sandface Flow Rate

In some cases the afterflow period can completely dominate the pressure buildup and the afterflow analysis is the only method of formation permeability and skin factor. A buildup pressure test with afterflow is considered as a variable flow rate test. In this case the Duhamel integral can be used to derive solutions for time-dependent boundary conditions.

It was assumed that the afterflow rate can be approximated by an exponential function where the exponent is expressed as quadratic equation of shut-in time. A direct method of calculating the Duhamel integral is presented (Kutasov 2015). This solution is utilized to estimate values of formation permeability and skin factor. An example of processing of field data for one case is shown.

During a buildup pressure test the closure is usually made at the wellhead. Due to compressibility of the fluid in the wellbore, sandface flow does not become zero instantly. This phenomenon is called "afterflow" or "wellbore storage". Several theoretical methods (Russell's equation, McKinley method) have been presented for analyzing the pressure response during the afterflow period in order to determine formation permeability and skin factor (Dake 1978). It is known, however, that the results obtained from these techniques are less accurate than those from the Horner analysis. As was mentioned by Dake (1978), in some cases (e.g., Middle East) the afterflow period can completely dominate the pressure buildup and the afterflow analysis is the only method of determining the essential reservoir parameters. A buildup pressure test with afterflow can be considered as a variable flow rate test. In this case the Duhamel integral (the superposition theorem) can be used to derive solutions for time-dependent boundary conditions from time-independent boundary conditions. Many investigators consider that the exponential function of time as a good approximation of the sandface rate during afterflow (e.g., Van Everdingen 1953; Hurst 1953; Kuchuk and Ayestaran 1985a). Kuchuk and Ayestaran (1985a) used the Laplace transform of the Duhamel integral (the Ramey's integration method) and obtained a long-term approximation of this integral. *We developed a direct method of calculating the above-mentioned integral.* Below we assume that afterflow bottom-hole pressure and sandface flow rate were recorded during a build-up test. In cases when the Middle-Time Region (infinite-acting reservoir) is very short and representative data (to apply Horner's technique) cannot be obtained, the solution of the Duhamel integral can be utilized to estimate approximate values of formation permeability and skin factor. Simple computer programs can be prepared to process field data. Below we present an example of processing of field data for one case.

9.1 Duhamel Integral

It is well known that the superposition theorem (Duhamel integral) can be used to derive solutions for time-dependent boundary conditions (Carslaw and Jaeger 1959). In our case:

$$p_{wtD} = \frac{p_i - p_{wt}(t)}{M} = \int_0^t q_D(\tau) \frac{d}{dt} p_{wD}(t-\tau) d\tau, \tag{9-1}$$

$$q = q_0 q_D, \quad q_D = 1, \qquad t < t_1,$$

$$q = q_0 q_D, \qquad q_D = \exp\left[-\alpha\Delta t - \beta(\Delta t)^2\right], \qquad \Delta t = t - t_1, \qquad \alpha > 0, \quad \beta > 0, \tag{9-2}$$

where q is the fluid flow rate, q_0 is the reference flow rate, q_D is the dimensionless fluid flow rate, α, β are coefficients, Δt is the shut-in time, and τ is the variable of integration.

It will be shown (in the field Example section) that Eq. (9-2) approximates the afterflow with better accuracy then the suggested expression (Kuchuk and Ayestaran 1985a):

$$q_D = \exp(-\alpha\Delta t). \tag{9-3}$$

In Eq. (9-1) the functions p_{wD} and dp_{wD}/dt are

$$p_{wD} = \frac{p_i - p_w(t)}{M} = -Ei\left(-\frac{D}{t}\right), \qquad \frac{d}{dt} p_{wD}(t-\tau) = \frac{\exp\left(-\frac{D}{t-\tau}\right)}{t-\tau}, \tag{9-4}$$

$$D = \frac{r_w^2}{4\eta}, \quad M = 70.6\frac{q_0\mu B_v}{kh}, \quad \eta = \frac{0.0002637k}{\phi c_t \mu}, \tag{9-5}$$

where μ is the viscosity, ϕ is the porosity, η is the hydraulic diffusivity, B_v is the oil formation volume factor, c_t is the total compressibility, D and M are some parameters, Ei is the exponential integral, h is the reservoir thickness, k is the formation permeability, r_w is the well radius, p_i is the initial reservoir pressure, p_{wt} is the bottom-hole pressure at a variable flow rate, and μ is the viscosity.

For q_D (Eqs. (9-2) and (9-3)) the integral (9-1) cannot be integrated readily and, therefore, we used the following approach: firstly, we obtained the solution of the Duhamel integral when q_D can be expressed by a polynomial of some degree n. Secondly, we will show that the function q_D (Eqs. (9-2) and (9-3)) can be approximated (with a good accuracy) by a second and fourth degree polynomials.

9.2 Approximation of q_D by a Polynomial

Let

$$q_D(t) = \sum_0^n a_i \Delta t^i, \qquad t > t_1, \tag{9-6}$$

where $a_0, a_1, a_2, \ldots\ldots a_n$ are constant coefficients.

By using conventional assumptions

$$\left\{ \begin{array}{l} t_D = \dfrac{\eta(t_1 + \Delta t)}{r_w^2} \gg 1, \qquad \Delta t_D = \dfrac{\eta \Delta t}{r_w^2} \gg 1, \\[3mm] \exp\left(-\dfrac{1}{4t_D}\right) \approx 1, \quad \exp\left(-\dfrac{1}{4\Delta t_D}\right) \approx 1. \end{array} \right\}. \tag{9-7}$$

The integration of Eq. (9-1) provides for $t > t_1$

$$p_{Dws} = \frac{p_i - p_{ws}}{M} = -Ei\left(-\frac{1}{4t_D}\right) - F_n(\Delta t, n) + Ei\left(-\frac{1}{4\Delta t_D}\right) - q_D Ei\left(-\frac{1}{4\Delta t_D}\right), \tag{9-8}$$

$$F_n(\Delta t, n) = \sum_{i=1}^n a_i \Delta t^i \sum_{j=1}^i \frac{1}{j}, \tag{9-9}$$

$$t_D = \frac{\eta(t_1 + \Delta t)}{r_w^2}, \qquad \Delta t_D = \frac{\eta \Delta t}{r_w^2}.$$

For $n \le 4$ a commercially available software Maple 7 (Maple 7 Learning Guide 2001) was utilized to verify Eq. (9-8). We should to note that earlier we obtained a similar equation for a drawdown variable flow rate (Kutasov 1987b). Taking into account the skin factor

$$p_{Dws} = \frac{p_i - p_{ws}}{M} = -Ei\left(-\frac{1}{4t_D}\right) - F_n(\Delta t) + Ei\left(-\frac{1}{4\Delta t_D}\right) - q_D Ei\left(-\frac{1}{4\Delta t_D}\right) + 2sq_D. \tag{9-10}$$

At $t = t_1$ the pressure drop is

$$p_i - p_{wf} = M\left[-Ei\left(-\frac{1}{4t_{pD}}\right) + 2s\right], \qquad t_{pD} = \frac{\eta t_1}{r_w^2}. \tag{9-11}$$

9.3 Modification of the e^{-x} Function

In order to use Eqs. (9-2) and (9-3) we have to approximate the exponential function with a polynomial of some degree n. In this section we will show that a quadratic expression suggested by Korn and Korn (1968) approximates the function q_D (Eqs. (9-2) and (9-3)) with a good accuracy

$$z = e^{-x} \approx 1 + ax + bx^2, \qquad 0 \le x \le \ln 2, \tag{9-12}$$

$$a = -0.9664, \qquad b = 0.3536, \qquad \Delta z_{max} = 0.003.$$

The last formula can be modified for values of $x > \ln 2$.

Let

$$x_1 = \ln 2, \qquad x_2 = \frac{x}{x_1}, \qquad J = Integer(x_2),$$

$$u = -Jx_1, \qquad e_1 = e^u.$$

Then

$$z_1 = \exp(-x) = e_1 \cdot \exp(-x - u), \qquad x + u \le \ln 2. \tag{9-13}$$

In Table 9-1 the functions exp(-x) and z_1 are compared. It can be seen that function z_1 approximates the exponential function with good accuracy.

Table 9-1. Comparison of $y = $ exp(-x) and z_1 functions, $R = \dfrac{z_1 - y}{y}$.

-x	y	z_1	$z_1 - y$	R, %
0.10	0.90484	0.90690	0.00206	0.23
0.20	0.81873	0.82086	0.00213	0.26
0.40	0.67032	0.67002	-0.00030	-0.05
0.60	0.54881	0.54746	-0.00136	-0.25
0.80	0.44933	0.45039	0.00106	0.24
1.00	0.36788	0.36838	0.00050	0.13
1.20	0.30119	0.30051	-0.00069	-0.23
1.40	0.24660	0.24671	0.00011	0.04
1.60	0.20190	0.20241	0.00051	0.25
1.80	0.16530	0.16518	-0.00012	-0.07
2.00	0.13534	0.13502	-0.00031	-0.23
2.50	0.08208	0.08201	-0.00007	-0.09
3.00	0.04979	0.04991	0.00012	0.24
3.50	0.03020	0.03023	0.00003	0.10
4.00	0.01832	0.01827	-0.00005	-0.25
4.50	0.01111	0.01112	0.00001	0.07
5.00	0.00674	0.00676	0.00002	0.27
6.00	0.00248	0.00248	0.00000	-0.15
7.00	0.00091	0.00091	0.00000	0.18
8.00	0.00034	0.00034	0.00000	0.00
9.00	0.00012	0.00012	0.00000	-0.04
10.00	0.00005	0.00005	0.00000	0.15

For

$$x = \alpha \Delta t + \beta (\Delta t)^2$$

we obtain from Eqs. (9-2) and (9-13).

$$q_D = e_1 \cdot \left[1 + a\left(\alpha \Delta t + \beta (\Delta t)^2 + u \right) + b\left(\alpha \Delta t + \beta (\Delta t)^2 + u^2 \right) \right]. \tag{9-14}$$

Thus we obtained a fourth degree polynomial

$$q_D = a_0 + a_1 \Delta t + a_2 (\Delta t)^2 + a_3 (\Delta t)^3 + a_4 (\Delta t)^4, \tag{9-15}$$

where

$$a_0 = e_1 \left(1 + au + bu^2 \right),$$

$$a_1 = e_1 \left(a\alpha + 2b\alpha u \right),$$

$$a_2 = e_1 \left(a\beta + b\alpha^2 + 2b\beta u \right),$$

$$a_3 = 2e_1 b\alpha\beta,$$

$$a_4 = e_1 b\beta^2.$$

9.4 Working Formulas

In this section we present working formulas for processing data obtained from a buildup test with afterflow.

9.4.1 Equation (9-2)

Let us consider two cases.

A. The value of the initial formation pressure is known.
B. The value of the initial formation pressure is not known.

From Eq. (9-9) we obtain

$$F_4 (\Delta t) = a_1 \Delta t + \frac{3}{2} a_2 \Delta t^2 + \frac{11}{6} a_3 \Delta t^3 + \frac{25}{12} a_4 \Delta t^4. \tag{9-16}$$

A.

For large values of t_D and Δt_D the logarithmic approximation of the $Ei(-x)$ function can be used,

$$-Ei(-x) \approx -\ln x - 0.57722, \tag{9-17}$$

where 0.57722 is the Euler's constant. The last equation for values of $x < 0.001$ approximates the $Ei(-x)$ function with an error less than 0.016% and with absolute accuracy of 0.001. Taking into account Eq. (9-17), Eq. (9-8) can be rewritten:

$$\frac{p_i - p_{ws}}{q_D} = M\left[\frac{1}{q_D}\ln\frac{t_1 + \Delta t}{\Delta t} - \frac{1}{q_D}F_4(\Delta t) + \ln(4t_{pD}) - 0.57722 + 2s + \ln\frac{\Delta t}{t_1}\right]. \quad (9\text{-}18)$$

Let

$$Y = \frac{p_i - p_{ws}}{q_D}, \quad X = \frac{1}{q_D}\ln\frac{t_p + \Delta t}{\Delta t} - \frac{1}{q_D}F_4(\Delta t) + \ln\frac{\Delta t}{t_1}. \quad (9\text{-}19)$$

$$B_1 = \ln(4t_{pD}) - 0.57722 + 2s = \ln t_{pD} + 0.80907 + 2s. \quad (9\text{-}20)$$

Then

$$Y = MX + MB_1 \quad (9\text{-}21)$$

and by using a linear regression program we can estimated the slope (M) and intercept (Int). The parameter B_1 is

$$B_1 = \frac{Int}{M}. \quad (9\text{-}22)$$

The formation permeability and the hydraulic diffusivity (η) are determined from Eq. (9-5). The skin factor is computed from Eq. (9-20):

$$s = \frac{1}{2}\left(B_1 - 0.80907 - \ln t_{pD}\right). \quad (9\text{-}23)$$

B.

From Eqs. (9-9) and (9-10) we obtain

$$-p_{ws} + p_{wf} = M\left[-Ei\left(-\frac{1}{4t_D}\right) - F_4(\Delta t) - Ei\left(-\frac{1}{4\Delta t_D}\right)(q_D - 1) + 2s(q_D - 1) + Ei\left(-\frac{1}{4t_{pD}}\right)\right]. \quad (9\text{-}24)$$

By using the logarithmic approximation of the $Ei(-x)$ function the last equation transforms to

$$-p_{ws} + p_{wf} = M\left[\ln\frac{t_1 + \Delta t}{t_1} - F_4(\Delta t) + \left(\ln\frac{t_1\eta}{r_w^2} + 0.80907 + \ln\frac{\Delta t}{t_1}\right)(q_D - 1) + 2s(q_D - 1)\right] \quad (9\text{-}25)$$

or

$$\frac{-p_{ws} + p_{wf}}{q_D - 1} = M\left[\frac{1}{q_D - 1}\ln\frac{t_1 + \Delta t}{t_1} - \frac{1}{q_D - 1}F_4(\Delta t) + \ln\frac{t_1\eta}{r_w^2} + 0.80907 + \ln\frac{\Delta t}{t_1} + 2s\right]. (9\text{-}26)$$

Let

$$Y = MX + MB, \quad (9\text{-}27)$$

where

$$B = \ln \frac{t_1 \eta}{r_w^2} + 0.80907 + 2s,$$

(9-28)

$$Y = \frac{p_{wf} - p_{ws}}{q_D - 1}, X = \frac{1}{q_D - 1} \ln \frac{t_1 + \Delta t}{t_1} - \frac{1}{q_D - 1} F_4 \left(\Delta t \right) + \ln \frac{\Delta t}{t_1}$$

(9-29)

and by using a linear regression program we can estimated the slope (*M*) and intercept (*Int*). The parameter *B* is

$$B = \frac{Int}{M}.$$

(9-30)

The formation permeability and the hydraulic diffusivity (η) are determined from Eq. (9-5).The skin factor is computed from Eq. (9-28):

$$s = \frac{1}{2} \left(B - 0.80907 - \ln \frac{t_1 \eta}{r_w^2} \right).$$

(9-31)

The initial reservoir pressure is estimated from Eq. (9-11).

9.4.2 Equation (9-3)

Following changes should be made to apply the derived working formulas when $\beta = 0$.

1. Eqs. (9-14) and (9-15) should be replaced by Eqs. (9-14)* and (9-15)*.

$$q_D = e_1 \cdot \left[1 + a\left(\alpha \Delta t \right)^2 + u + b\left(\alpha \Delta t + u \right)^2 \right],$$

(9-14)*

$$q_D = a_0 + a_1 \Delta t + a_2 \left(\Delta t \right)^2,$$

(9-15)*

where

$$a_0 = e_1 \left(1 + au + bu^2 \right),$$

$$a_1 = e_1 \left(a\alpha + 2b\alpha u \right),$$

$$a_2 = e_1 b\alpha^2.$$

2. Eq. (9-16) should be replaced by Eq. (9-16)*.

$$F_2 \left(\Delta t \right) = a_1 \Delta t + \frac{3}{2} a_2 \Delta t^2.$$

(9-16)*

9.5 Field Example

The following example is an 8-hour buildup test in an oil well (Meunier et al. 1985). Well, fluid, and rock properties are presented in Table 9-2. Pressure and sandface rate were recorded simultaneously every 3 seconds with a wireline pressure/flowmeter combination tool. The semilog slope was first determined from the points after 1 hour, when the afterflow becomes negligible (Figure 9-1).

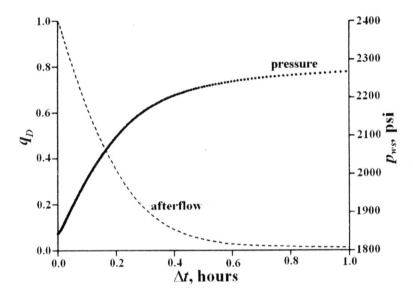

Figure 9-1. Buildup test data—pressure and dimensionless sandface rate (Meunier et al. 1985).

The Authors (Meunier et al. 1985) determined the following parameters:

$p_i = p^* = 2.405$ psia, $k = 403.2$ md, and $s = 1.55$.

Below we will use the early (0.107–0.2449 hr) recorded pressure and flow rate data.

Table 9-2. Well, rock and fluid properties.

h	Formation thickness, ft	100
q_o	Production flow rate, STB/D	9,200
r_w	Well radius, ft	0.1875
μ	Viscosity, cP	1.24
c_t	Compressibility, 1/psia	0.000011
B	Formation volume factor, RB/STB	1.24
ϕ	Formation porosity, fraction	0.27
t_p	Production time, hrs	158.0
p_{wf}	Flowing pressure, psia	1844.65

9.6 Processing of Field Data and Results

Five techniques were used to process field data (Table 9-3). Eqs. (9-2), (9-3), and Eq. (9-6) at $n = 2$ were utilized to approximate the measured afterflow rate. Eqs. (9-2) and (9-3) can be transformed to

$$\log q_D = -\alpha\Delta t - \beta\Delta t^2,$$ (9-32)

$$\log q_D = -\alpha\Delta t.$$ (9-33)

Now by using linear and quadratic regression computer programs the coefficients in Eqs. (9-3), (9-33) and Eq. (9-6) ($n = 2$) can be estimated (Table 9-3).

Table 9-3. Results of calculations.

CaseNo.	$-\alpha$, 1/hr	$-\beta$, 1/(hr)^2	k, md	S	$p,^*$ psia
1 Eq. (9-2)	4.21074	5.45388	446.45	2.96	2413.11
2 Eq. (9-2)	4.21074	5.45388	431.61	2.27	2405.00
3 Eq. (9-3)	5.17181	0	243.38	-0.93	2542.66
4 Eq. (9-3)	5.17181	0	460.15	3.37	2405.00
5 Eq. (9-6) ($n = 2$)	$a_0 = 1.0023$ $a_1 = -4.3404$ hr^{-1} $a_2 = 5.3053$ hr^{-2}		430.51	2.24	2405.00
Meunier et al. 1985	-	-	403.2	1.55	2405.00

Table 9-4. Measured (q^*) and calculated (q) values of the flow rate, $R = \dfrac{q^* - q}{q^*}100$.

Δt, hrs	q^*, STB/D	Cases 1 and 2 q, STB/D	R, %	Cases 3 and 4 q, STB/D	R, %	Case 5 q, STB/D	R,%
0.0107	8790.1	8789.2	0.01	8704.7	0.97	8799.5	-0.11
0.0136	8687.8	8679.2	0.10	8575.1	1.30	8687.1	0.01
0.0166	8564.9	8566.0	-0.01	8443.1	1.42	8571.7	-0.08
0.0224	8346.0	8349.0	-0.04	8193.6	1.83	8351.2	-0.06
0.0282	8130.6	8134.6	-0.05	7951.5	2.20	8133.9	-0.04
0.0341	7914.6	7919.1	-0.06	7712.5	2.55	7916.2	-0.02
0.0399	7700.6	7710.0	-0.12	7484.6	2.80	7705.6	-0.06
0.0457	7496.9	7503.6	-0.09	7263.4	3.11	7498.2	-0.02
0.0545	7195.5	7196.0	-0.01	6940.3	3.55	7189.9	0.08
0.0632	6890.2	6898.5	-0.12	6634.9	3.71	6892.4	-0.03
0.0720	6605.2	6604.5	0.01	6339.7	4.02	6599.1	0.09
0.0807	6396.0	6321.0	1.17	6060.8	5.24	6316.5	1.24
0.0903	5984.5	6016.5	-0.53	5767.2	3.63	6013.3	-0.48
0.1102	5390.8	5413.8	-0.43	5203.2	3.48	5413.4	-0.42
0.1310	4817.2	4825.9	-0.18	4672.5	3.00	4827.7	-0.22
0.1557	4184.7	4184.5	0.01	4112.2	1.73	4187.0	-0.06
0.1807	3585.9	3597.5	-0.32	3613.4	-0.77	3599.2	-0.37
0.2107	2976.5	2973.9	0.09	3094.1	-3.95	2974.4	0.07
0.2449	2378.6	2365.3	0.56	2592.5	-8.99	2369.2	0.39

In the first case Eq. (9-2) was used to approximate the flow rate and it was assumed that the initial pressure is not known. In the second case the value of $p_i = 2405$ psi were taken from Meunier et al. (1985). Similarly, for cases 3 and 4 Eq. (9-3) was used to approximate measured values of q. In case 5 Eq. (9-6) ($n = 2$) was utilized to approximate measured values of q. In all cases the regression method was used to obtain the coefficients in Eqs. (9-2), (9-3) and Eq. (9-6) ($n = 2$). The accuracy of approximation can be seen from Table 9-4. The results of calculations are presented in Tables 9-3 and 9-5. We should to remember that values of $k = 403.2$ md and $s = 1.55$ were estimated from the Horner plot (large values of shut-in time).

Table 9-5. Measured ($p_{ws}*$) and calculated values of p_{ws} for 5 cases (Table 9-3) compared.

Δt, hrs	$p_{ws}*$, psia	$p_{ws}* - p_{ws}$, psia				
		1	2	3	4	5
0.0107	1859.07	−0.57	−4.68	0.46	5.56	−4.27
0.0136	1863.35	−0.56	−4.43	0.34	4.69	−4.08
0.0166	1867.81	−0.55	−4.17	0.18	3.83	−3.88
0.0224	1876.97	−0.06	−3.21	0.31	2.77	−3.04
0.0282	1885.99	0.24	−2.44	0.26	1.72	−2.36
0.0341	1894.87	0.26	−1.96	−0.06	0.53	−1.96
0.0399	1903.80	0.50	−1.28	−0.11	−0.26	−1.35
0.0457	1912.72	0.76	−0.58	−0.08	−0.86	−0.70
0.0545	1925.94	0.95	0.25	−0.12	−1.70	0.08
0.0632	1938.79	1.07	0.99	−0.10	−2.28	0.79
0.0720	1951.53	1.11	1.65	−0.01	−2.67	1.43
0.0807	1963.92	1.16	2.27	0.20	−2.80	2.06
0.0903	1977.10	0.96	2.71	0.31	−2.93	2.50
0.1102	2003.45	0.52	3.48	0.85	−2.52	3.32
0.1310	2028.89	−0.62	3.52	1.07	−1.98	3.41
0.1557	2056.41	−3.55	1.82	0.01	−1.79	2.74
0.1807	2081.78	−5.30	1.23	0.64	−0.38	1.76
0.2107	2108.50	−7.86	−0.12	0.68	0.92	−0.10
0.2449	2134.43	−11.18	−2.28	−0.33	1.67	−2.61

9.7 Discussion of Results

From Tables 9-3 and 9-5 follows that Eq. (9-2) approximates the after production flow rate better than Eq. (9-3). Application of Eq. (9-3) when the initial reservoir pressure is not known (case 3) does not allow to calculate the values of the formation permeability and skin factor with a satisfactory accuracy (Table 9-3). At the same time the agreement between measured and calculated values of bottom-hole shut-in pressure is very good (Table 9-5). However, when the initial reservoir pressure is known Eq. (9-2) provides satisfactory estimates of the formation permeability and skin factor (Tables 9-3 and 9-5). As can be observed from Tables 9-2 and 9-4, Eq. (9-2) allows determine the formation permeability and skin factor in both cases (at a given or unknown value of p_i) with a good accuracy. It is interesting to note that for this field example cases 2 and 5 provide practically the same values of the formation permeability and skin factor. This shows, indirectly, that Eq. (9-12) is a valid approximation of the exponential function.

Final Deductions

A novel approach to obtain solution of the Duhamel integral, when the afterflow rate is an exponential function of the shut-in time, is suggested. Working formulas for determination of formation permeability and skin factor are developed. Our study confirms the well-known conclusions that the results (determination of formation permeability and skin factor) obtained from the afterflow data are less accurate than those from the Horner analysis.

10

Analyzing the Pressure Response during the Afterflow Period: Determination of the Formation Permeability and Skin Factor

As we mentioned in Chapter 9, in some cases the afterflow period can completely dominate the pressure buildup and the afterflow analysis is the only method of formation permeability and skin factor. Assuming that the afterflow rate can be approximated by an exponential function we developed a direct method of computing the Duhamel integral (see Chapter 9). In the previous Chapter we utilized the flow rate-time and pressure-time data to present a data processing technique for estimation formation permeability and skin factor.

However in most of pressure buildup tests only pressure-time data are available. The presented below (for this case) solution enables to reconstruct of the sandface flow rate during the afterflow period and is utilized to estimate values of formation permeability and skin factor. Three examples of processing of field data are presented. Simple computer programs were prepared to process the field data. Below we present three examples of calculations.

10.1 Duhamel Integral

It is well known that the superposition theorem (Duhamel integral) can be used to derive solutions for time-dependent boundary conditions (Carslaw and Jaeger 1959). In our case:

$$p_{wtD} = \frac{p_t - p_{wt}(t)}{M} = \int_0^t q_D(\tau) \frac{d}{dt} p_{wD}(t - \tau) d\tau, \qquad (10\text{-}1)$$

$$\begin{cases} q = q_o q_D, & q_D = 1, & t < t_1, \\ q = q_o q_D, & q_D = e^{-\alpha \Delta t}, & \Delta t > 0, \quad \Delta t = t - t_1, \quad \alpha > 0 \end{cases},$$ (10-2)

where $p_{wD}(t_D)$ is the dimensionless sandface pressure for the constant-rate case without the skin effect, p_i is the initial reservoir pressure, p_{wt} is the bottom-hole pressure at a variable flow rate, τ is the variable of integration, t is time, q_0 is the reference flow rate, q_D is the dimensionless flow rate, and t_1 is the production period.

$$p_{wD} = \frac{p_i - p_w(t)}{M} = -Ei\left(-\frac{D}{t}\right),$$ (10-3)

$$\frac{d}{dt} p_{wD}(t-\tau) = \frac{\exp\left(-\dfrac{D}{t-\tau}\right)}{t-\tau},$$ (10-4)

$$D = \frac{r_w^2}{4\eta}, \quad M = 70.6\frac{q_o \mu B_v}{kh}, \quad \eta = \frac{0.0002637k}{\phi c_t \mu},$$ (10-5)

where the following parameters: well radius (r_w), viscosity (μ), permeability (k), total compressibility (c_t), reference flow rate (q_0) are expressed in oilfield units, and porosity (φ) is a fraction. For $q_D = \exp(-\alpha\Delta t)$ the integral (10-1) cannot be integrated readily and, therefore, we used the following approach: firstly, we obtained the solution of the Duhamel integral when q_D can be expressed by a polynomial of some degree n. Secondly, we will show that the function $\exp(-\alpha\Delta t)$ can be approximated (with a good accuracy) by a quadratic equation ($n = 2$).

10.2 Approximation of q_D by a Polynomial

Let

$$q_D(t) = \sum_0^n a_i \Delta t^i, \quad t > t_1,$$ (10-6)

where $a_0, a_1, a_2, \ldots\ldots a_n$ are constant coefficients.

By using conventional assumptions

$$\begin{cases} t_D = \dfrac{\eta(t_1 + \Delta t)}{r_w^2} \gg 1, \quad \Delta t_D = \dfrac{\eta\Delta t}{r_w^2} \gg 1, \\ \\ \exp\left(-\dfrac{1}{4t_D}\right) \approx 1, \quad \exp\left(-\dfrac{1}{4\Delta t_D}\right) \approx 1. \end{cases}$$ (10-7)

The integration of Eq. (10.1) provides for $t > t_1$:

$$p_{Dws} = \frac{p_i - p_{ws}}{M} = -Ei\left(-\frac{1}{4t_D}\right) - F_n(\Delta t) + Ei\left(-\frac{1}{4\Delta t_D}\right) - q_D Ei\left(-\frac{1}{4\Delta t_D}\right), \tag{10-8}$$

$$F_n(\Delta t) = \sum_{i=1}^{n} a_i \Delta t_i \sum_{j=1}^{i} \frac{1}{j}. \tag{10-9}$$

For $n \leq 4$ and a commercially available software Maple (Maple 7 Learning Guide 2001) was utilized to verify Eq. (10-8). We should to note that earlier we obtained a similar equation for a drawdown variable flow rate (Kutasov 1987). Taking into account the skin factor

$$p_{Dws} = \frac{p_i - p_{ws}}{M} = -Ei\left(-\frac{1}{4t_D}\right) - F_n(\Delta t, n) + Ei\left(-\frac{1}{4\Delta t_D}\right) - q_D Ei\left(-\frac{1}{4\Delta t_D}\right) + 2sq_D \tag{10-10}$$

and using the logarithmic approximation of the *Ei* function we obtained

$$-Ei\left(-\frac{1}{4\Delta t_D}\right) + 2s \approx \ln\left(4t_{pD} \frac{\Delta t}{t_p}\right) - 0.57722 + 2s = U + \gamma, \tag{10-11}$$

$$U = \ln\left(4t_{pD}\right) - 0.57722 + 2s, \qquad t_{pD} = \frac{\eta t_1}{r_w^2}, \qquad \gamma = \ln\frac{\Delta t}{t_1}, \tag{10-12}$$

$$p_i - p_{ws} = M\left[\ln\frac{t_p + \Delta t}{\Delta t} - F_n(\Delta t, n) + q_D(U + \gamma)\right], \tag{10-13}$$

At $t = t_1$ the pressure drop is

$$p_i - p_{wf} = M\left[-Ei\left(-\frac{1}{4t_{pD}}\right) + 2s\right] \approx MU. \tag{10-14}$$

In order to use Eqs. (10-8) and (10-9) we have to approximate the exponential function with a polynomial of some degree n. It was shown that a quadratic expression approximates the $q_D = e^{-\alpha \Delta t}$ function with a good accuracy (Eqs. (9-14)*, (9-15)* and (9-16*)).

10.3 Working Formulas

Taking into account Eq. (9-16)* and from Eqs. (10-13) and (10-14) we obtain

$$y = \frac{p_i - p_{wf}}{p_i - p_{ws}} = \frac{U}{\ln\dfrac{t_1 + \Delta t}{\Delta t} - F_2(\Delta t, \alpha) + q_D(U + \gamma)}, \tag{10-15}$$

or

$$U(\Delta t, \alpha) = \frac{y\left(\ln\dfrac{t_1 + \Delta t}{\Delta t} - F_2(\Delta t, \alpha) + q_D \gamma\right)}{1 - yq_D}. \tag{10-16}$$

Let as assume that two values of shut-in pressures are available: p_{ws1} at Δt_1 and p_{ws2} at Δt_2, then from Eq. (10-16) we can derive an equation for computing the value of α

$$U(\Delta t = \Delta t_1, \alpha) = U(\Delta t = \Delta t_2, \alpha). \tag{10-17}$$

To use a computer program the last equation should be rewritten as

$$U(\Delta t = \Delta t_1, \alpha) - U(\Delta t = \Delta t_2, \alpha) = \varepsilon, \tag{10-18}$$

where ε is a small value and depends on the accuracy of y ratio. The Newton method was used (Grossman 1977) for solving Eq. (10-18). In this method a solution of an equation is sought by defining a sequence of numbers which become successively closer and closer to the solution. The conditions, which guarantee that Newton method in our case will work and provide a unique solution, are satisfied (Grossman 1977). In the subroutine which utilizes the Newton method the following parameters were used: (a) the starting value of t_{Dap} is 0.1; (b) the time increment is 1.1; (c) the absolute accuracy of the ratio of y is $\varepsilon = 0.0001$. When N records of p_{ws} and Δt are available, then $N(N-1)/2$ values of α can be computed and its squared averaged value is estimated.

Now Eq. (10-10) transforms to

$$\frac{p_i - p_{ws}}{q_D} = M\left[\frac{1}{q_D}\ln\frac{t_1 + \Delta t}{\Delta t} - \frac{1}{q_D}F_2(\Delta t, \alpha) + \ln(4t_{pD}) - 0.57722 + 2s + \ln\frac{\Delta t}{t_1}\right]. \tag{10-19}$$

Let

$$Y = \frac{p_i - p_{ws}}{q_D}, \qquad X = \frac{1}{q_D}\ln\frac{t_p + \Delta t}{\Delta t} - \frac{1}{q_D}F_2(\Delta t, \alpha) + \ln\frac{\Delta t}{t_1}. \tag{10-20}$$

$$B_1 = \ln(4t_{pD}) - 0.57722 + 2s. \tag{10-21}$$

Then,

$$Y = M(X + B_1)$$

and by using a linear regression program we can estimated the slope (M) and intercept (Int). The parameter B_1 is

$$B_1 = \frac{Int}{M}. \tag{10-22}$$

The formation permeability and the hydraulic diffusivity (η) are determined from Eq. (10-5). The skin factor is computed from Eq. (10-21).

$$s = \frac{1}{2}\left(B_1 - 0.80907 - \ln t_{pD}\right).$$

(10-23)

10.4 Field Examples

Below we present three field examples. Only shut-in time and build-up pressures data taken during the after-production period (Table 10-1) were utilized to estimate the coefficient α and to determine values of formation permeability and skin factor. Well, fluid, and rock properties are presented in Table 10-2.

Table 10-1. Input data.

Case 1		Case 2		Case 3	
Δt, hrs	$p_{ws,}$ psia	Δt, hrs	$p_{ws,}$ psia	Δt, hrs	$p_{ws,}$ psia
0.15	3680	0.10	3057	0.50	4675
0.20	3723	0.21	3153	0.66	4705
0.30	3800	0.31	3234	1.00	4733
0.40	3866	0.52	3249	1.50	4750
0.50	3920			2.00	4757
1.00	4103				
2.00	4250				

Table 10-2. Well, rock and fluid properties.

Symbol	Field cases	1	2	3
h	Formation thickness, ft	69.0	482	20.0
q_o	Production flow rate, STB/D	250	4900	123
r_w	Well radius, ft	0.198	0.354	0.300
μ	Viscosity, cP	0.8	0.2	1.0
c_t	Compressibility, 1/psia	0.000017	0.0000226	0.000020
B	Formation volume factor, RB/STB	1.136	1.55	1.22
ϕ	Formation porosity, fraction	0.039	0.09	0.20
t_p	Production time, hrs	13,630	310	97.6
p_{wf}	Flowing pressure, psia	3,534	2761	4506

Example 1(Lee 1982): The following example is a 72-hour buildup test in an oil well. It seems that this field example was adapted (and modified) from Matthews and Russell (1967). The well was produced for an effective time of 13,630 hours at the final rate ($t = t_p = 13,630$ hours). The flow rate history prior to the well shut-in is not known and this can be a reason why the quadratic average deviation for the coefficient α is too large (2.36 ± 0.78 1/hr, Table 10-4). We should mention that for the 0.15–2.0 hour interval (Table 10-2) we used 17 pairs of time—pressure data. It was estimated that after 6 hours the afterflow ceases to distort the pressure buildup data. The semilog slope was determined from the points after 6 hours, when the afterflow becomes negligible. The following parameters were determined (Lee 1982; Matthews and Russell 1967):

$p_i = p^* = 4,577$ psia, $k = 7.65$ md, and $s = 6.37$.

At processing of field data we used the value of $p_i = 4,577$ psia. The results of calculations are presented in Table 10-3.

Table 10-3. Results of calculations and comparison of results.

Case No.	This study			Δt, hrs	Conventional technique		
	k, md	s	α, 1/hr		k, md	s	α,1/hr
1	6.50	6.51	2.36 ± 0.78	0.15 – 2.0	7.65	6.37	6.0 – 72.0
2	12.52	7.30	10.61 ± 1.96	0.1 – 0.52	12.80	8.60	0.8 – 37.5
3	50.35	5.09	2.64 ± 0.23	0.5 – 2.0	50.00	6.00	2.5 – 12.0

The afterflow flow rate at $\Delta t = 6$ hours is $q = 250 \cdot \exp(-2.36 \cdot 6) = 0.00018 (STB/D)$.

Example 2 (Earlougher 1977): The following example is a 37.5-hour buildup test in an oil well. It was estimated that after 0.75 hours the afterflow cease to distort the pressure buildup data. Well, fluid, and rock properties are presented in Table 10-3. The semilog slope was determined from the points after 0.75 hours, when the afterflow becomes negligible. The following parameters were determined:

$p_i = p^* = 3365.7$ psia, $k = 12.8$ md, and $s = 8.6$.

At processing of field data we used the value of $p_i = 3365.7$ psia.

The input data and the results of calculations are presented in Tables (10-3)–(10-5).

Table 10-4. Estimated values of α.

Δt_{s1}, hr	Δt_{s2}, hr	p_{ws1}, psi	p_{ws2}, psi	α, 1/hr
0.10	0.31	3057	3234	9.936
0.10	0.52	3057	3249	10.633
0.21	0.52	3153	3249	8.240
0.31	0.52	3234	3249	13.651

Table 10-5. Comparison of observed (p_{ws}) and calculated values of shut-in pressure (p^*_{ws}).

t_s, hr	Q, STB/D	p_{ws}, psia	p^*_{ws}, psia	$p_{ws} - p^*_{ws}$, psia
0.1000	1695.1	3057.0	3074.8	−17.8
0.2100	527.4	3153.0	3185.2	−32.2
0.3100	182.4	3234.0	3223.1	10.9
0.5200	19.6	3249.0	3249.5	−0.5

The afterflow flow rate at $\Delta t = 0.75$ hours is

$$q = 4,900 \cdot \exp(-10.61 \cdot 0.75) = 4,900 \cdot \exp(-7.96) = 1.72 \, (\text{STB/D}).$$

Example 3 (Dake 1978): The following example is a 97.6-hour buildup test in a discovery well. It was estimated that after of about 2 hours the afterflow cease to distort the pressure buildup data. Well, fluid, and rock properties are presented in Table 10-3. The semilog slope was determined from the points after 2.5 hours. The following parameters were determined:

$p_i = p^* = 4,800$ psia, $k = 50.0$ md, and $s = 6.0$.

At processing of field data we use the value of $p_i = 4,800$ psia.

The results of calculations are presented in Tables 10-4, 10-6 and 10-7.

Table 10-6. Calculated values of α, Case 3.

Δt_1, hr	Δt_2, hr	p_{ws1}, psi	p_{ws2}, psi	α, 1/hr
0.50	2.0	4675	4757	2.58
0.50	1.5	4675	4750	2.59
0.50	1.0	4675	4733	2.50
0.66	1.5	4705	4757	2.45
0.66	2.0	4705	4750	3.09

Table 10-7. Afterflow flow rate, measured (p_s^*) and calculated (p_{ws}) buildup pressures, Case 3.

t_s, hr	Q, STB/D	p_{ws}, psia	p^*_{ws}, psia	$p_{ws} - p^*_{ws}$, psia
0.50	32.8	4681	4675	-6
0.66	21.5	4704	4705	1
1.00	8.7	4732	4733	1
1.50	2.3	4750	4750	0
2.00	0.6	4757	4757	0

The afterflow flow rate at $\Delta t = 2.5$ hours is

$$q = 123 \cdot \exp(-2.64 \cdot 2.5) = 123 \cdot \exp(-6.6) = 0.17 \, (\text{STB/D}).$$

We can conclude that the calculated values of the formation permeability and skin factor are in a satisfactory agreement with those calculated by the conventional Horner method (Table 10-4). The dimensionless flow rates for three field cases versus shut-in time are presented in Figure 10-1.

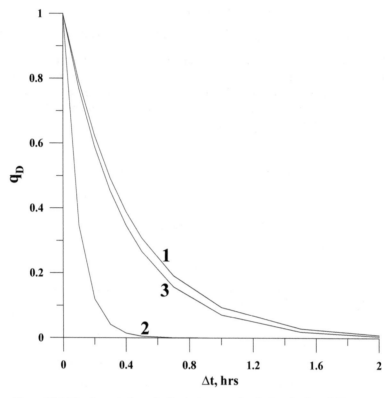

Figure 10-1. The dimensionless afterflow rate versus shut-in time for three field cases.

Thus, a novel approach to obtain solution of the Duhamel integral, when the afterflow rate is an exponential function of the shut-in time, is suggested. Working formulas for determination of formation permeability and skin factor are developed.

11

Application of the Horner Method for a Well Produced at a Constant Bottom-hole Pressure

11.1 Horner Method: A Short Description

The Horner method is widely used to process the pressure-buildup test data for wells produced at a constant flow rate. When the flow rate before shut-in changes relatively slowly, the real production time should be replaced by the Horner corrected (adjusted) production time and the last established flow rate should be used. When the flow rate during the test varies within wide limits, the theory of superposition is used to model the production history. Below an approximate equation for the pressure drop in the reservoir at the time $t = t_p$ (production time) is used as initial pressure distribution for the shut-in period. To determine the buildup pressures, we used the solution of the diffusivity equation that describes the pressure change on the axis of a cylindrical body with a known initial pressure distribution placed in the infinite medium at a constant initial pressure (Kutasov 1989). Because we did not use the principle of superposition, the equations obtained are valid for practically all values of production or shut-in time.

Assumptions: It is assumed that the flow of the single-phase liquid in the reservoir is described with a diffusivity equation in the cylindrical coordinates:

$$\frac{\partial^2 p}{\partial r^2} + \frac{1}{r}\frac{\partial p}{\partial r} = \frac{\phi \mu c}{k}\frac{\partial p}{\partial t}, \qquad (11\text{-}1)$$

where φ is the porosity, μ is the viscosity, c is the compressibility, and k is the permeability. The use of Eq. (11-1) implies a number of assumptions: isothermal flow of fluids of small and constant compressibility, constant porosity, permeability, and fluid viscosity, and neglect of the gravity forces. It is also assumed that effective wellbore radius can be used (Uraiet and Raghavan 1980).

Dimensionless parameters: The dimensionless flowing time based on the effective wellbore radius and the dimensionless radial distance are defined by

$$t_D = \frac{ktp}{\phi c \mu r_{wa}^2}, \qquad r_D = \frac{r}{r_{wa}}, \tag{11-2}$$

where r_{wa} is the apparent wellbore radius.

$$r_{wa} = r_w e^{-s}, \tag{11-3}$$

where s is the skin factor.

The dimensionless Horner corrected (adjusted) production time is defined by

$$t_D^* = Gt_D. \tag{11-4}$$

The function G was defined earlier (see Chapter 2).

The dimensionless pressure drop during production is defined by

$$p_D(r_D, t_D) = \frac{p_i - p(t, t_p)}{p_i - p_{wf}}, \tag{11-5}$$

where $p(r, t_p)$ is the pressure in the reservoir at a distance r and time t_p. The flowing bottom-hole pressure (BHP) p_{wf}, and the initial reservoir pressure, p_i are assumed to be constant.

The dimensionless BHP after shut-in is defined by

$$p_{sD} = \frac{p_i - p_{ws}}{p_i - p_{wf}}, \tag{11-6}$$

where $p_{ws}(\Delta t)$ is the BHP after shut-in and Δt is the shut-in time.

11.2 Pressure Drop

To determine the function (r, t_p) it is necessary to obtain the solution of Eq. (11-1) under the following initial and boundary conditions:

$$p(r, 0) = p_i, \qquad r_{wa} \leq r \leq \infty, \tag{11-7a}$$

$$p(r_{wa}, t) = p_{wf}, \qquad p(\infty, t) = p_i, \qquad t > 0. \tag{11-7b}$$

It is well known that for a well producing at a constant BHP with the initial condition as in Eq. (11-7a), the diffusivity equation has a solution in a complex integral form (Carslaw and Jager 1959). Using the results of a numerical solution, we have found (Kutasov 1999) that for values of dimensionless production time $t_D > 1$. Equation (11-8) can be used to approximate the pressure distribution in the reservoir during production:

$$p_D = \frac{p_i - p(r,t)}{p_i - p_{wf}} = \frac{Ei\left(-\dfrac{r_D^2}{4t_D^*}\right)}{Ei\left(-\dfrac{1}{4t_D^*}\right)}, \qquad r_D \geq 1, \tag{11-8}$$

where Ei is the exponential integral. Thus, for values of $tp \gg 1$ (assuming also that the effective radius concept can be used), the expression for the flowing pressure (at $r_D = 1$) can be approximated by

$$p_i - p_{wf} = \frac{q_L \mu}{4\pi kh} = \left[-Ei\left(-\frac{1}{4t_D^*}\right)\right], \tag{11-9}$$

where q_L is the last production rate (at $t = t_p$), and h is the reservoir thickness.

To determine the temperature in the wellbore ($r = 0$) after the production period, we used the solution of the diffusivity equation that describes cooling along the axis of a cylindrical body with known initial temperature distribution, placed in an infinite medium of constant temperature (Carslaw and Jaeger 1959). We obtained (Kutasov 1989)

$$p_{sD}(\Delta t) = \frac{p_i - p_{ws}(\Delta t)}{p_i - p_{wf}} = 1 - \frac{Ei\left(-\dfrac{1}{4t_D^*} - \dfrac{1}{4\Delta t_D}\right)}{Ei\left(-\dfrac{1}{4t_D^*}\right)}, \tag{11-10}$$

where Δt_D is the dimensionless shut-in time. Combining Eqs. (11-6), (11-9) and (11-10), we obtain

$$p_i - p_{ws}(\Delta t) = \frac{q_L \mu}{4\pi kh}\left[-Ei\left(-\frac{1}{4t_D^*}\right) + Ei\left(-\frac{1}{4t_D^*} - \frac{1}{4\Delta t_D}\right)\right]. \tag{11-11}$$

The logarithmic approximation of the Ei function is valid for the argument

$$\frac{1}{4t_D^*} + \frac{1}{4\Delta t_D} < 0.01. \tag{11-12}$$

Using the log approximation for the Ei function in Eqs. (11-9) and (11-11), we can express values of p_{wf}, p_{ws} and s in oilfield units as

$$p_{wf} = p_i - m\left(\log\frac{kt_p^*}{\varphi c \mu r_w^2} - 3.2275 + 0.8686s\right), \tag{11-13}$$

$$m = \frac{162.6 q_L B \mu}{kh}, \tag{11-14}$$

$$p_{wf} = p_i - m\left(\log\frac{t_p^* + \Delta t}{\Delta t}\right), \tag{11-15}$$

$$s = 1.1513 \left[\left\{ \frac{\left[p_{ws}\left(\Delta t = 1\right) - p_{wf} \right]}{m} \right\} - \log \frac{k}{\phi c \mu r_w^2} + \log \frac{t_p^* + 1}{t_p^*} + 3.2275 \right]. \qquad (11\text{-}16)$$

Thus, we obtained the working equations for the modified Horner method, where the last flow and adjusted production time are used (Earlougher 1977, p. 40).

Below we present a simulated example. The input data are taken from Earlougher (1977, Example 4.4). Assume that prior to the build-up test the well was produced for 100 hours and $k = 6.5$ md, $\mu = 1.35$ cP, $h = 190$ ft, $B = 1$ RB/STB, $\varphi c = 2.05 \ 10^{-6}$ psi^{-1}, $p_i - p_{wf} = 1{,}000$ psi, $r_w = 1$ ft, $s = 0$.

Step 1. Calculation of the dimensionless production time:

$$t_{Da} = 0.0002637 \frac{kt_p}{\phi c \mu r_w^2} = 0.0002637 \frac{6.5 \cdot 100}{1.35 \cdot 2.05 \cdot 10^{-6} \cdot 1^2} = 61935.$$

Step 2. From Eq. (2-2) we find the value of the G–function:

$$G\left(t_{Da} = 61935\right) = 1.10.$$

Step 3. Thus the adjusted production time is:

$100 \cdot 1.10 = 110$ (hours).

Step 4. From Eq. (5-5) we find the value of the dimensionless flow rate:

$$q_D \left(t_D = 61935\right) = 0.1674.$$

Step 5. And finally, the last value of the production rate is:

$$q = \frac{kh\left(p_i - p_{wf}\right)}{141.2 B \mu} q_D = \frac{6.5 \cdot 190}{141.2 \cdot 1.35 \cdot 1 \cdot 10^{-6} \cdot 1^2} 0.1674 = 1{,}087 \text{(STB/D)}.$$

And the Horner plot is:

$$p_{wf} = p_i - m \left(\log \frac{110 + \Delta t}{\Delta t} \right).$$

Now let us assume that the well was fractured and the skin factor was reduced from $s = 0$ to $s = -4.0$.

Then

Step 1.

$$r_{wa} = 1 \cdot e^{+4} = 54.6, \quad t_D = 0.0002637 \frac{kt_p}{\phi c \mu r_{wa}^2} = 0.0002637 \frac{6.5 \cdot 100}{1.35 \cdot 2.05 \cdot 10^{-6} \cdot 54.6^2} = 20.78.$$

Step 2.

$$G\left(t_{Da} = 20.78\right) = 1.324.$$

Step 3. Thus the adjusted production time is:

$100 \cdot 1.324 = 132.4 \ (\text{hours})$.

Step 4. From Eq. (5-5) find the value of the dimensionless flow rate:

$q_D \left(tDa = 20.78\right) = 0.4637$.

Step 5. Finally, the last value of the production rate is:

$$q = \frac{kh\left(p_i - p_{wf}\right)}{141.2B\mu} q_D = \frac{6.5 \cdot 190}{141.2 \cdot 1.35 \cdot 1 \cdot 10^{-6} \cdot 1^2} 0.4637 = 3.011 \left(\text{STB/D}\right).$$

And the Horner plot is

$$p_{wf} = p_i - m\left(\log \frac{132.4 + \Delta t}{\Delta t}\right).$$

We can conclude that an analytical solution is obtained for buildup-pressure predictions for a well produced at constant BHP in an infinite reservoir. This solution is valid for practically any values of production or shut-in times. For large dimensionless production times, the obtained equation can be simplified and the Horner equation can be derived. In many cases, the Horner buildup method using the last flow rate and the adjusted production time can be applied to estimate the formation permeability, skin factor, and initial reservoir pressure.

|12|

Step-Pressure Testing

12.1 Two Attractive Features

During a step-pressure test fluid is produced at two successive constant pressure flow periods. Figure 12-1 illustrates pressure of a step-pressure test. Either a decreasing or increasing pressure sequence may be used. Similarly, two-rate tests have two attractive features.

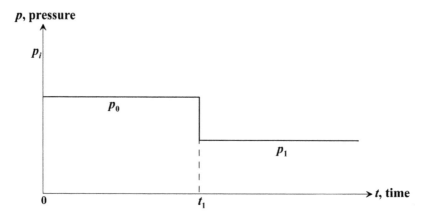

Figure 12-1. A step-pressure test.

1. The tests do not require that the well be shut in if it is already producing at a constant bottom-hole pressure.
2. The wellbore storage effects on the test data are short-lived. Sengul (1983) presented a technique of analyzing step-pressure test data for damaged wells. A general equation is presented below that can be used without any limitations for processing step-pressure test data in fractured/stimulated or damaged wells. A computer program was also developed to speed up data processing (Kutasov 1998).

12.2 Working Equations

If one assumes that a well is producing against a constant pressure from an infinite-acting reservoir and the effective well bore radius concept can be used. In this case, flow rate in oil field units is defined by Eqs. (12-1) and (12-2).

$$q(t) = \frac{kh(p_i - p_{wf})}{141.2 \, \mu B} q_D(t_D), t_D = \frac{0.0002637 kt}{\varphi \, \mu c_t r_{wa}^2},$$ (12-1)

$$r_{wa} = r_w e^{-s},$$ (12-2)

where q is the flow rate, STB/D, k is the permeability, md, h is the reservoir thickness, ft, p_i is the initial reservoir pressure, psi, p_{wf} is the well flowing pressure, psi, μ is the viscosity, cP, B is the oil formation volume factor, RB/STB, q_D is the dimensionless flow rate, c_t is the compressibility, 1/psi, r_w is the well radius, ft, r_{wa} is the effective well radius, ft, φ is the porosity, s is the skin factor, and t_D is the dimensionless time.

For fractured wells the values of t_D can be very small. For large values of t_D a simple equation for the dimensionless flow rate can be used (Sengul 1983). A semi-theoretical Eq. (12-3) can approximate the dimensionless flow rate (see Chapter 5 and Table 5-1). Hence, the principle of superposition can be used without any limitations in this case.

$$q_D = \frac{1}{\ln(1 + D\sqrt{t_D})}, \quad D = d + \frac{1}{\sqrt{t_D} + b}, \quad a = 1.5708, \quad b = 4.9589.$$ (12-3)

Application of the principle of superposition obtains the flow rate for the second flow period, Eq. (12-4).

$$q(t) = \frac{kh(p_i - p_0)}{141.2 \, \mu B} \left[q_D(t_D) + \frac{(p_0 - p_1)}{(p_i - p_0)} q_D(\Delta t_D) \right],$$ (12-4)

$$t_{D1} = t_D \frac{t_1}{t}, \quad \Delta t_D = t_D \frac{\Delta t}{t}, \quad \Delta t = t - t_1.$$

During the second flow period, it is assumed that two flow rates were obtained $q_a = q(t = t_a)$ and $q_b = q(t = t_b)$. From these two equations, one can determine the formation permeability and wellbore skin factor. Equation (12-5) can be obtained from Eq. (12-4). From Eq. (12-5) we can estimate dimensionless time t_{D1}.

$$\frac{q(t_a)}{q(t_b)} = \frac{q_D(t_{Da}) + \gamma q_D(\Delta t_{Da})}{q_D(t_{Db}) + \gamma q_D(\Delta t_{Db})},$$ (12-5)

$$\gamma = \frac{p_0 - p_1}{p_i - p_0}, \quad t_{Da} = t_{D1} \frac{t_a}{t_1}, \quad t_{Db} = t_{D1} \frac{t_b}{t_1},$$

$$\Delta t_{Da} = t_{D1} \left(\frac{t_a}{t_1} - 1 \right), \quad \Delta t_{Db} = t_{D1} \left(\frac{t_b}{t_1} - 1 \right).$$

To use a computer program the last equation should be rewritten as

$$\frac{q(t_a)}{q(t_b)} - \frac{q_D(t_{Da}) + \gamma q_D(\Delta t_{Da})}{q_D(t_{Db}) + \gamma q_D(\Delta t_{Db})} = \varepsilon, \tag{12-6}$$

where ε is a small value and depends on the accuracy of the $q(t_a)/q(t_b)$ ratio.

Grossman (1977) has applied the Newton method for solving Eq. (12-6). In this method a solution of an equation is sought by defining a sequence of numbers which become successively closer and closer to the solution. The conditions, which guarantee that the Newton method in our case will work and provide a unique solution, are satisfied (Grossman 1977). In the subroutine which utilizes the Newton method the following parameters were used: (a) the starting value of t_{D1} is 2; (b) the time increment is 2.0; (c) the absolute accuracy of the ratio of y is $\varepsilon = 0.000001$ (Kutasov 1998). Next, from Eq. (12-4) (using q_a or q_b) formation permeability is calculated. Finally, the parameter t_{D1} is used to compute the apparent well bore radius. From Eq. (12-2) the well bore skin factor is calculated. Note that if N records of flow rates are available, then it is possible to obtain N $(N-1)/2$ values of k and s. In this case the regression technique can be used to analyze test data.

Example

Table 12-1 lists the test values Δt and q (Sengul 1983) observed in the oil well and computed related parameters. It was assumed that $k = 50$ md and $s = 0.0$. The program "Step" was used to estimate the formation permeability and skin factor (Table 12-2). The calculated k and s agree well with the assumed values.

Table 12-1. Step-pressure test data.

Sengul (1983)		
Δt, hr	q, STB/D	parameters
7.01	6,091	$h = 100$ ft
10.05	5,989	$\varphi = 0.15$
13.09	5,918	$c_t = 25 \cdot 10^{-6}$ 1/psi
16.02	5,865	$B = 1.2$ RB/STB
20.0	5,809	$\mu = 2$ cP
26.07	5,744	$t_1 = 1500$ hr
34.98	5,675	$p_i = 5,000$ psi
45.00	5,617	$p_0 = 3,500$ psi
55.00	5,573	$p_1 = 2,000$ psi
70.00	5,521	$r_w = 0.35$ ft
100.0	5,446	
140.0	5,377	
175.0	5,333	

Table 12-2. Results of calculations.

Δt_a, hr	Δt_b, hr	q_a, STB/D	q_b, STB/D	k, md	s
7.01	13.09	6091.0	5918.0	50.00	−0.05
10.05	16.02	5989.0	5865.0	49.93	−0.06
13.09	20.00	5918.0	5809.0	49.86	−0.07
16.02	26.07	5865.0	5744.0	49.91	−0.06
20.00	34.98	5809.0	5675.0	49.95	−0.06
26.07	45.00	5744.0	5617.0	49.77	−0.08
34.98	55.00	5675.0	5573.0	49.82	−0.08
45.00	70.00	5617.0	5521.0	50.15	−0.03
55.00	100.00	5573.0	5446.0	49.85	−0.07
70.00	140.00	5521.0	5377.0	49.63	−0.10
100.00	175.00	5446.0	5333.0	49.85	−0.07
140.00	175.00	5377.0	5333.0	50.38	0.01
7.01	175.00	6091.0	5333.0	49.89	−0.06

References to Part I

Bejan, A. 2004. Convection Heat Transfer. Wiley and Sons, NJ.

Bourdarot, G. 1998. Well Testing: Interpretation Methods. Editions Technip, Paris.

Bourdet, D., Ayoub, J.A. and Pirard, Y.-M. 1989. Use of the pressure derivative in well test interpretation. SPE Formation Evaluation, June, pp. 293–302. http://www.spe.org/publications/journals.php.

Carslaw, H.S. and Jaeger, J.C. 1959. Conduction of Heat in Solids. 2nd edition. Oxford Univ. Press, London.

Dake, L.P. 1978. Fundamentals of Reservoir Engineering. Elsevier, Amsterdam-London-N.Y.-Tokio.

Earlougher, R.C., Jr. 1977. Advances in Well Test Analysis. SPE New York, Dallas.

Edwards, L.M., Chilingar, G.V., Rieke, H.H. and Fertl, W.H. 1982. Handbook of Geothermal Energy. Gulf Publishing Co., Houston.

Edwardson, M.L., Girner, H.M., Parkinson, H.R., Williams, C.D. and Matthews, C.S. 1962. Calculation of formation temperature disturbances caused by mud circulation. Jour. of Petrol. Tech., pp. 416–426. http://www.spe.org/jpt/.

Ehlig-Economides, C.A. and Ramey, H.J. 1981. Pressure buildup for wells produced at constant pressure. SPEJ, February, pp. 105–114. http://www.spe.org/publications/journals.php.

Elder, J.W. 1981. Geothermal Systems. Academic Press, NY, San Francisco, London.

Eppelbaum, L.V., Modelevsky, M.M. and Pilchin, A.N. 1996. Thermal investigation in petroleum geology: the experience of implication in the Dead Sea Rift zone. Journal of Petroleum Geology, 19, No. 4, Israel, 425–444.

Eppelbaum, L.V. and Kutasov, I.M. 2006a. Temperature and pressure drawdown well testing: Similarities and differences. Jour. of Geophysics and Engineering, 3, No. 1, IOP Publishing, Bristol, pp. 12–20.

Eppelbaum, L.V. and Kutasov, I.M. 2006b. Determination of formation temperatures from temperature logs in deep boreholes: comparison of three methods. Journal of Geophysics and Engineering, 3, No. 4, IOP Publishing, Bristol, pp. 348–355.

Eppelbaum, L.V. and Kutasov, I.M. 2011. Estimation of the effect of thermal convection and casing on temperature regime of boreholes—a review. Journal of Geophysics and Engineering, 8, IOP Publishing, Bristol, R1–R10.

Eppelbaum, L.V. and Kutasov, I.M. 2013. Cylindrical probe with a variable heat flow rate: A new method for determination of the formation thermal conductivity. Central European Journal of Geosciences, 5, No. 4, Springer, New York, pp. 570–575.

Eppelbaum, L.V., Kutasov, I.M. and Pilchin, A.N. 2014. Applied Geothermics. Springer, New York.

Gradshtein, I.S. and Ryzhik, I.M. 1965. Table of Integrals, Series and Products. Oxford University Press, London, England.

Gretener, P.E. 1981. Geothermics: Using Temperature in Hydrocarbon Exploration. AAPG, Short Course Notes 17, American Association of Petroleum Geologists. http://www.aapg.org/.

Grossman, S.I. 1977. Calculus Academic Press NY, San Francisco, London.

Hawkins, M.F., Jr. 1956. A note on the skin effect. Trans. AIME, 207, American Institute of Mining, Metallurgical, and Petroleum Engineers, New York, pp. 356–357.

Horne, R.N. 1995. Modern Well Test Analysis: A Computer-Aided Approach. Petroway Inc., Palo Alto, CA.

Hurst, W. 1953. Establishment of the skin effect and its impediment to fluid flow into a well bore. Petrol. Engin., B6–B16I, Hindawi Publishing Corporation, Cairo. http://www.hindawi.com/journals/jpe/.

Jacob, C.E. and Lohman, S.W. 1952. Non-steady flow to a well of constant drawdown in an extensive aquifer. Trans. Amer. Geophys. Un., John Wiley and Sons, NJ, pp. 559–564.

Johnson, P.W. 1986. The relationship between radius of drainage and cumulative production. SPE 16035. http://www.spe.org/publications/journals.php.

Jorden, J.R. and Campbell, F.L. 1984. Well logging I—rock properties, borehole environment, mud and temperature logging. SPE of AIME, N.Y., Dallas, pp. 131–146.

Kappelmeyer, O. and Haenel, R. 1974. Geothermics with Special Reference to Application. Gebruder Borntrager. Berlin.

Korn, G.A. and Korn, T.M. 1968. Mathematical Handbook for Scientists and Engineers. 2nd Edition, McGraw-Hill Book Company, N.Y.

Kuchuk, F. and Ayestaran, L. 1985a. Analysis of simultaneously measured pressure and sandface flow rate in transient well testing. Jour. of Petroleum Technology, February, SPE, pp. 323–334. http://www.spe.org/publications/journals.php.

Kuchuk, F. and Ayestaran, L. 1985b. Author's reply to discussion of analysis of simultaneously measured pressure and sandface flow rate in transient well testing. Jour. of Petroleum Technology, October, SPE. http://www.spe.org/publications/journals.php.

Kutasov, I.M. 1987a. Dimensionless temperature, cumulative heat flow and heat flow rate for a well with a constant bore-face temperature. Geothermics, 16, No. 2, Elsevier, pp. 467–472.

Kutasov, I.M. 1987b. Pressure analysis for a variable flow rate drawdown test. SPE paper 17265.

Kutasov, I.M. 1989. Application of the Horner method for a well produced at a constant bottomhole pressure. Formation Evaluation, 3, No. 3, SPE, pp. 90–92.

Kutasov, I.M. 1998. Program Analyzes Step-Pressure Data. Oil & Gas Journal, Jan. 5, pp. 43–46.

Kutasov, I.M. 1999. Applied Geothermics for Petroleum Engineers. Elsevier.

Kutasov, I.M. 2003. Dimensionless temperature at the wall of an infinite long cylindrical source with a constant heat flow rate. Geothermics, 32, Elsevier, pp. 63–68.

Kutasov, I.M. 2007. Determination of calcium chloride brine concentration required to provide pressure overbalance. J. of Petrol. Science and Engineering, 58 Elsevier, pp. 133–137.

Kutasov, I.M. 2013. Short-Term Testing Method for Stimulated Wells—Field Examples. Jour. of Canadian Petroleum Technology, 52, No. 6, SPE, pp. 1–7.

Kutasov, I.M. 2015. Analyzing the pressure response during the afterflow period—Determination of the formation permeability and skin factor. Proceed. of the World Geothermal Congress, Melbourne, Australia, 1–9.

Kutasov, I.M. and Eppelbaum, L.V. 2003. Prediction of formation temperatures in permafrost regions from temperature logs in deep wells—field cases. Permafrost and Periglacial Processes, 14, No. 3, John Wiley and Sons, N.J., pp. 247–258. www.interscience.wiley.com. DOI: 10.1002/ppp.457.

Kutasov, I.M. and Eppelbaum, L.V. 2005. Drawdown test for a stimulated well produced at a constant bottomhole pressure. First Break 23(2): 25–28. http://onlinelibrary.wiley.com/journal/10.1111/%28 ISSN%291365-2397/issues.

Kutasov, I.M. and Eppelbaum, L.V. 2007. Temperature well testing—utilization of the Slider's method. Jour. of Geophysics and Engineering 4(1): 1–6.

Kutasov, I.M. and Eppelbaum, L.V. 2008. Designing an interference well test in a geothermal reservoir. Proceedings. 32rd Workshop on Geothermal Reservoir Engineering, Stanford University, SGP-TR-185, Stanford, California, January 28–30.

Kutasov, I.M. and Eppelbaum, L.V. 2009. Estimation of the geothermal gradients from single temperature log-field cases. Jour. of Geophysics and Engineering 6(2): 131–135.

Kutasov, I.M. and Eppelbaum, L.V. 2010. A new method for determination of formation temperature from bottom-hole temperature logs. Jour. of Petroleum and Gas Engineering 1(1): 1–8. http://www.academicjournals.org/journal/JPGE.

Kutasov, I.M. and Eppelbaum, L.V. 2011. Recovery of the thermal equilibrium in deep and superdeep wells: Utilization of measurements while drilling data. Proceed. of the 2011 Stanford Geothermal Workshop, Stanford, USA, SGP-TR-191, pp. 1–7.

Kutasov, I.M. and Eppelbaum, L.V. 2012a. New method evaluated efficiency of wellbore stimulation. Oil & Gas Journal 110(8): 22–24.

Kutasov, I.M. and Eppelbaum, L.V. 2012b. Geothermal investigations in permafrost regions—the duration of temperature monitoring after wellbores shut-in. Geomaterials 2(4): 82–93.

Kutasov, I.M. and Eppelbaum, L.V. 2013a. Cementing of casing—temperature increase at cement hydration. Proceed. of the 2013 Stanford Geothermal Workshop, Stanford, USA, pp. 1–6.

Kutasov, I.M. and Eppelbaum, L.V. 2013b. Optimization of temperature observational well selection. Exploration Geophysics 44(3): 192–198.

Kutasov, I.M. and Eppelbaum, L.V. 2013c. Cementing of casing: the optimal time lapse to conduct a temperature log. Oil Gas European Magazine 39(4): 190–193.

Kutasov, I.M. and Eppelbaum, L.V. 2014. Temperature regime of boreholes: Cementing of production liners. Proceed. of the 2014 Stanford Geothermal Workshop, Stanford, USA, pp. 1–5.

Kutasov, I.M. and Eppelbaum, L.V. 2015. Wellbore and Formation Temperatures during Drilling, Shut-in and Cementing of Casing. Proceed. of the World Geothermal Congress, Melbourne, Australia, pp. 1–12.

Kutasov, I.M., Eppelbaum, L.V. and Kogan, M. 2008. Interference well testing—variable fluid flow rate. Jour. of Geophysics and Engineering 5(1): 86–91.

Kutasov, I.M. and Kagan, M. 2003a. Cylindrical probe with a constant temperature—determination of the formation thermal conductivity and contact thermal resistance. Geothermics 32: 187–193.

Kutasov, I.M. and Kagan, M. 2003b. Determination of the skin factor for a well produced at a constant Bottomhole Pressure. Jour. of Energy Resources Techn., 125, March, pp. 61–63.

Kutasov, I.M. and Hejri, C. 1984. Drainage radius of a well produced at constant bottomhole pressure in an infinite acting reservoir. SPE 13382. http://www.spe.org/publications/journals.php.

Kutasov, I.M. and Water, F.V. 1989. Factors at higher pressures and temperatures. Oil & Gas Journal, March 1989, pp. 102–104.

Lee, J. Well Testing. 1982. SPE Monograph Series, Texas. http://www.spe.org/publications/journals.php.

Waterloo Maple Inc. 2001. Maple 7 Learning Guide, Waterloo, Canada.

Matthews, C.S. and Russell, D.G. 1967. Pressure Buildup and Flow Tests in Wells, Vol. 1. SPE, Henry L. Doherty Series, Texas. http://www.spe.org/publications/journals.php.

McDonald, S.W. 1983. Evaluation of production tests in oil wells stimulated by massive acid fracturing offshore Qatar. Jour. of Petrol. Technol., March, pp. 496–506. http://www.spe.org/publications/journals.php.

Meunier, D., Wittmann, M.J. and Stewart, G. 1985. Interpretation of pressure buildup test using *in situ* measurement of afterflow. Jour. of Petroleum Technology, January, pp. 143–152. http://www.spe.org/publications/journals.php.

Muskat, M. 1946. Flow of Homogeneous Fluids Through Porous Media. J. W. Edwards Inc., Ann Arbor, MI.

Prats, M. 1982. Thermal Recovery. Monograph Series 7, Society of Petroleum Engineers, Dallas.

Ramey, H.J. (Jr). 1962. Wellbore heat transmission. Jour. of Petroleum Technology 14(4): 427–435. http://www.spe.org/publications/journals.php.

Sabet, M.A., 1991. Well Test Analysis. Contrib. in Petroleum Geology and Engineering. Houston, Gulf Publishing.

Serra, O. 1984. Fundamentals of Well Log Interpretation, Vols. 1 and 2. Elsevier.

Schechter, R.S. 1992. Oil Well Stimulation. Prentice Hall, Englewood Cliffs, NJ.

Sengul, M.M. 1983. Analysis of step-pressure tests. SPE Paper 12175 presented at the 58th Annual Technical Conf. and Exhibition, San Francisco, California, 5–8 October, 1983. http://www.spe.org/publications/journals.php.

Smith, L.P. 1937. Heat flow in an infinite solid bounded internally by a cylinder. Jour. Appl. Phys., Vol. 8, American Institute of Physics, pp. 441–448.

Somerton, W.H. 1992. Thermal Properties and Temperature Related Behavior of Rock/Fluid Systems. Developments in Petroleum Science. Elsevier. http://www.journals.elsevier.com.

Tittman, J. 1986. Geophysical Well Logging. Academic Press, NY, San Francisco, London.

Van Everdingen, A.F. 1953. The skin effect and its influence on the productive capacity of a well. Trans., AIME, 198 American Institute of Mining, Metallurgical, and Petroleum Engineers, New York, pp. 171–176.

Van Everdingen, A.F. and Hurst, W. 1949. The application of the Laplace transformation to flow problems in reservoirs. Trans., AIME, Vol. 186, American Institute of Mining, Metallurgical, and Petroleum Engineers, New York, pp. 305–324.

Vosteen, H.-D. and Schellschmidt, R. 2003. Influence of temperature on thermal conductivity, thermal capacity and thermal diffusivity for different types of rock. Physics and Chemistry of the Earth, 28, Cambridge University Press, Cambridge, pp. 499–509.

Uraiet, A.A. and Raghavan, R. 1980. Pressure buildup for wells at a constant bottomhole. Jour. of Petroleum Technology, October, pp. 1813–1824. http://www.spe.org/publications/journals.php.

PART II

Temperature Well Testing

13

Determination of Formation Temperature from Bottom-Hole Temperature Logs: A Generalized Horner Method

A new technique has been developed for determination of the formation temperature from bottom-hole temperature logs. The adjusted circulation time concept, and a semi-analytical equation for the dimensionless temperature at the wall of an infinite long cylindrical source with a constant heat flow rate, is used to obtain the working formula. It is shown that the transient shut-in temperature is a function of the mud circulation and shut-in time, formation temperature, thermal diffusivity of formations, and well radius. The sensitivity of the predicted values of formation temperature to the thermal diffusivity is shown. Two examples of calculations are presented.

The determination of physical properties of reservoir fluids, calculation of hydrocarbon volumes (estimation of oil and gas formation volume factors, and gas solubility), predictions of the gas hydrate prone zones, well log interpretation, determination of heat flow density, and evaluation of geothermal energy resources require knowledge of the undisturbed formation temperature.

In most cases bottom-hole temperature surveys are mainly used to determine the temperature of the Earth's interior. The drilling process, however, greatly alters the temperature of formation immediately surrounding the well. The temperature change is affected by the duration of drilling fluid circulation, the temperature difference between the reservoir and the drilling fluid, the well radius, the thermal diffusivity of the reservoir, and the drilling technology used. Given these factors, the exact determination of formation temperature at any depth requires a certain length of time in which the well is not in operation. In theory, this shut-in time is infinitely long to reach the original condition. There is, however, a practical limit to the time required for the difference in temperature between the well wall and surrounding reservoir to become vanishingly small.

The objective of this chapter is to suggest a new approach in utilizing bottom-hole temperature logs in deep wells and to present a working formula for determining the undisturbed formation temperature. For this reason we do not here conduct a review and analysis of relevant publications. We will discuss only the Horner method, which is often used in processing field data. Earlier we used the condition of material balance to describe the pressure build-up for wells produced at constant bottom-hole pressure (Kutasov 1989b). The build-up pressure equation was derived on the basis of an initial condition approximating the pressure profile in the wellbore and in the reservoir at the time of shut-in. It was shown that a modified Horner method (Kutasov and Eppelbaum 2005b) could be used to estimate the initial reservoir pressure and formation permeability.

For many cases, the shut-in time (the time since end of mud circulation, Δt) or the duration of the mud circulation period (t_c) cannot be estimated (with a sufficient accuracy) from well reports. In these cases methods of correcting the bottom-hole temperature data for the uncertainty in determining values of Δt and t_c can be utilized (Majorowicz et al. 1990; Waples and Ramly 2001; Waples et al. 2004).

In this chapter we will consider only bottom-hole temperature logs. This means that the thermal disturbance of formations (near the well's bottom) is caused by short drilling time and, mainly, by one (prior to logging) continuous drilling fluid circulation period. The duration of this period is usually 3–24 hours. It is known that the same differential diffusivity equation describes the transient flow of incompressible fluid in porous medium and heat conduction in solids. As a result, a correspondence exists between the following parameters: volumetric flow rate, pressure gradient, mobility (formation permeability and viscosity ratio), hydraulic diffusivity coefficient; and heat flow rate, temperature gradient, thermal conductivity and thermal diffusivity. Thus, the same analytical solutions of the diffusivity equation (at corresponding initial and boundary conditions) can be utilized for determination of the above-mentioned parameters. In this study we will use a similar technique (Kutasov 1989a) for determination undisturbed (initial) formation temperature from bottom-hole temperature logs. As will be shown below, by introducing the adjusted circulation time concept, a new method of determining static formation temperature can be developed.

13.1 Mathematical Models

The determination of static formation temperatures from well logs requires knowledge of the temperature disturbance produced by circulating drilling mud.

To determine the temperature distribution $T(r, t)$ in formations we will consider three mathematical models to describe the thermal effect of the circulating drilling fluid.

13.1.1 Constant bore-face temperature

The results of field and analytical investigations have shown that in many cases the temperature of the circulating fluid at a given depth can be assumed constant during drilling or production (Lachenbruch and Brewer 1959; Ramey 1962; Edwardson et al. 1962; Jaeger 1961; Kutasov et al. 1966; Raymond 1969). In this case it is necessary to obtain a solution of the diffusivity equation for the following boundary and initial conditions:

$$\begin{cases} T(r,0) = T_i, & r_w \leq r < \infty, \\ T(r_w,t) = T_w, & T(\infty,t) = T_i, \quad t > 0 \end{cases}, \tag{13-1}$$

where T_w is the wall temperature, T_i is the initial (undisturbed) formation temperature, r is the radial coordinate and r_w is the radius of the borehole.

It is known that in this case the diffusivity equation has a solution in complex integral form (Jaeger 1956; Carslaw and Jaeger 1959). Jaeger (1956) presented results of a numerical solution for the dimensionless temperature $T_D(r_D, t_D)$ with values of r_D ranging from 1.1 to 100 and t_D ranging from 0.001 to 1000.

The dimensionless temperature T_D, dimensionless distance r_D, and dimensionless time t_D are:

$$T_D(r_D,t_D) = \frac{T(r,t) - T_i}{T_w - T_i}, \qquad r_D = \frac{r}{r_w}, \qquad t_D = \frac{\chi t}{r_w^2}. \tag{13-2}$$

where χ is the thermal diffusivity of formations.

Lachenbruch and Brewer (1959) have shown that the wellbore shut-in temperature mainly depends on the amount of thermal energy transferred to (or from) formations during drilling. For this reason we present below formulas which allow us to calculate the heat flow rate (q) and cumulative heat flow (Q) from the wellbore per unit of length:

$$q = 2\pi\lambda(T_w - T_i)q_D(t_D), \tag{13-3}$$

where λ is the thermal conductivity of formations and q_D is the dimensionless heat flow rate. Analytical expressions for the function $q_D = f(t_D)$ are available only for asymptotic cases or for large values of t_D. We have found (Kutasov 1987) that for any values of dimensionless production time a semi-theoretical (13-4) can be used to forecast the dimensionless heat flow rate (see Chapter 5):

$$q_D = \frac{1}{\ln\left(1 + D\sqrt{t_D}\right)}, \tag{13-4}$$

$$D = d + \frac{1}{\sqrt{t_D} + b}, \quad d = \frac{\pi}{2}, \quad b = \frac{2}{2\sqrt{\pi} - \pi}. \tag{13-5}$$

The cumulative heat flow from (or into) the wellbore per unit of length is given by:

$$Q = 2\pi\rho c_p r_w^2 (T_w - T_i)Q_D(t_D), \tag{13-6}$$

where c_p is the heat capacity of formations, ρ is the density of formation, and $Q_D(t_D)$ is the dimensionless cumulative heat flow (Kutasov 1987).

13.1.2 Cylindrical source with a constant heat flow rate

In this case the transient temperature T_w is a function of time, thermal conductivity, and volumetric heat capacity of formations. Analytical expression for the function T_w is available only for large values of the dimensionless time (t_D). To determine the temperature T_w it is necessary to obtain the solution of the diffusivity equation under the following boundary and initial conditions:

$$T(t=0,r)=T_i, \qquad r_w \leq r < \infty, \tag{13-7}$$

$$\left(r\frac{\partial T}{\partial r}\right)_{r_w} = -\frac{q}{2\pi\lambda}, \qquad T(t,r\to\infty)\to T_i, \qquad t>0. \tag{13-8}$$

It is well-known that in this case the diffusivity equation has a solution in complex integral form (Van Everdingen and Hurst 1949; Carslaw and Jaeger 1959). Chatas (Lee 1982) tabulated this integral for $r = r_w$ over a wide range of values of t_D.

For the wall transient temperature we obtained the following semi-analytical equation (Kutasov 2003)

$$T_w = T(t,r_w) = T_i + \frac{q}{2\pi\lambda}\ln\left[1+\left(c-\frac{1}{a+\sqrt{t_D}}\right)\sqrt{t_D}\right], \tag{13-9}$$

$$a = 2.7010505, \qquad c = 1.4986055.$$

Let us introduce the dimensionless wall temperature

$$T_{wD}(t_D) = \frac{2\pi\lambda(T_w - T_i)}{q}. \tag{13-10}$$

Then

$$T_{wD}(t_D) = \ln\left[1+\left(c-\frac{1}{a+\sqrt{t_D}}\right)\sqrt{t_D}\right]. \tag{13-11}$$

Values of T_{wD} calculated from Eq. (13-11) and results of a numerical solution ("exact" solution) by Chatas (Lee 1982) were compared (Kutasov 2003). The agreement between values of T_D calculated by these two methods was very good. For this reason the principle of superposition can be used without any limitations.

13.1.3 Well as a linear source

It is clear from physical considerations that, for large values of dimensionless time, the solutions for cylindrical and linear sources should converge. To develop the solution for a linear source, the boundary condition expressed by Eq. (13-8) should be replaced by the condition

$$\lim_{r\to 0}\left(r\frac{\partial T}{\partial r}\right) = -\frac{q}{2\pi\lambda} \qquad t>0. \tag{13-12}$$

and the well-known solution for the infinitely long linear source with a constant heat flux rate in an infinite-acting medium is (Carslaw and Jaeger 1959)

$$T_r(r,t) = T_i - \frac{q}{4\pi\lambda} Ei\left(-\frac{r^2}{4\chi t}\right),$$ (13-13)

where $Ei(-x)$ is the exponential integral.

Introducing the dimensionless radial temperature

$$T_{rD}\left(r_D, t_D\right) = \frac{T(r,t) - T_i}{T_w - T_i}$$ (13-14)

we obtain

$$T_{wD}\left(t_D\right) = -\frac{1}{2} Ei\left(-\frac{1}{4t_D}\right).$$ (13-15)

The radial temperature distribution can be expressed by the following equation

$$T_{rD}\left(r_D, t_D\right) = \frac{Ei\left(-\dfrac{r_D^2}{4t_D}\right)}{Ei\left(-\dfrac{1}{4t_D}\right)}.$$ (13-16)

Earlier it was shown (see Chapter 2) that by using the adjusted circulation time concept a well with a constant borehole wall temperature can be substituted by a linear source with a constant heat flow rate.

13.2 Circulation Period

Field investigations have shown that the bottom-hole circulating (without penetration) fluid temperature after some stabilization time can be considered constant (Figures 13-1 and 13-2). The solid curves in Figure 13-1 present the calculated circulating mud temperatures (at a constant heat transfer coefficient) by using the Raymond (1969) model.

We also found that the exponential integral can be used to describe the temperature field of formations around a well with a constant bore-face temperature (Kutasov 1987)

$$T_{rD}\left(r_D, t_D\right) = \frac{T(r,t) - T_i}{T_w - T_i} = \frac{Ei\left(-\dfrac{r_D^2}{4t_{cD}^*}\right)}{Ei\left(-\dfrac{1}{4t_{cD}^*}\right)}, \qquad t_{cD}^* = \frac{\chi t_c^*}{r_w^2},$$ (13-17)

where t_s^* is the adjusted time of the "thermal disturbance".

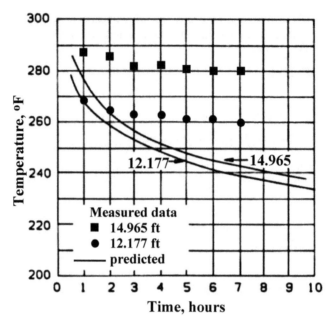

Figure 13-1. Comparison of measured and predicted circulating mud temperatures, Well 1 (after Sump and Williams 1973).

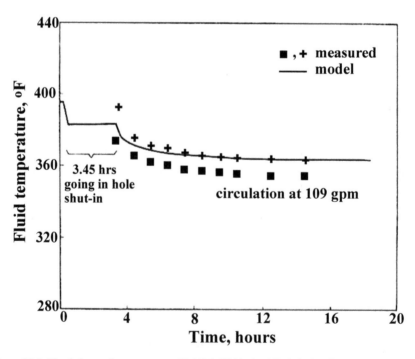

Figure 13-2. Circulating mud temperature at 23,669 ft (7214 m)—Mississippi well (Wooley et al. 1984). Courtesy of Society of Petroleum Engineers.

In Table 13-1 values of $T_{rD}(r_D, t_D)$ computed with Eq. (13-17) and results of a numerical solution are compared. The agreement between values of $T_D(r_D, t_D)$ calculated by these two methods is seen to be good. It is easy to see that Eqs. (13-16) and (13-17) are similar and identical at $G \rightarrow 1$ (see Eqs. (2-1) and (2-2)).

Table 13-1. Dimensionless radial temperature $T_{rD}(r_D, t_D) \cdot 1000$ for a well with constant bore-face temperature, first line is obtained from Eq. (13-17), second line is the numerical solution (Jaeger 1956).

t_D/r_D	1.1	1.2	1.5	2.0	3.0	5.0	7.0	10.0
2.0	912	834	642	418	172	22	1	0
	924	854	677	458	194	22	1	0
5.0	934	875	726	543	310	97	26	2
	940	886	746	568	332	101	24	2
10.0	945	896	771	614	404	180	77	18
	949	903	784	631	422	188	77	16
20.0	953	912	804	668	481	266	148	59
	956	916	813	681	497	277	153	57
30.0	957	919	820	694	520	314	194	93
	959	922	827	705	534	325	201	94
50.0	961	926	837	723	564	370	253	144
	963	929	843	731	574	381	260	146

Introducing the adjusted circulation time into Eq. (13-9) we obtain

$$T_w = T(t, r_w) = T_i + \frac{q}{2\pi\lambda} \ln\left[1 + \left(c - \frac{1}{a + \sqrt{Gt_D}} \right) \sqrt{Gt_D} \right], \tag{13-18}$$

where $q = q(t = t_c)$, the last heat flow rate.

13.3 Horner Method

The Horner method is widely used in petroleum reservoir engineering and in hydrogeological exploration to process the pressure-build-up test data for wells produced at a constant flow rate. From a simple semilog linear plot the initial reservoir pressure and formation permeability can be estimated. Using the similarity between the transient response of pressure and temperature build-up, it was suggested to use the Horner method for prediction of formation temperature from bottom-hole temperature surveys (Timko and Fertl 1972; Dowdle and Cobb 1975; Fertl and Wichmann 1977; Jorden and Campbell 1984). It is assumed that the wellbore can be considered as a linear source of heat.

Santoyo et al. (2000) performed an interesting thermal evolution study of the LV-3 well in the Tres Virgenes geothermal field, Mexico. Several series of temperature logs were run during LV-3 drilling and shut-in operations. The temperature build-up tests were limited to short shut-in times (up to 24 hours). Static formation temperatures

(SFT) were computed by five analytical methods (including the Horner plot), which are the most commonly used in the geothermal industry. The authors observed that the SFT predictions made by use of the Horner method were always less than the temperatures provided by other methods. In the Horner method the thermal effect of drilling is approximated by a constant linear heat source. This energy source is in operation for some time t_c and represents the time elapsed since the drill bit first reached the given depth. For a continuous drilling period the value of t_c is identical with the duration of mud circulation at a given depth. The well-known expression for the borehole temperature is (Eq. (13-13) $r = r_w$)

$$T_w\left(r_w,t_c\right)-T_i = -\frac{q}{4\pi\lambda}Ei\left(-\frac{1}{4t_{cD}}\right). \tag{13-19}$$

Using the principle of superposition the following equation for shut-in temperature can be obtained:

$$T_s\left(r_w,t_s\right)-T_i = \frac{q}{4\pi\lambda}\left[-Ei\left(-\frac{1}{4(t_{cD}+t_{sD})}\right)+Ei\left(-\frac{1}{4t_{sD}}\right)\right], \qquad t_{sD} = \frac{\chi\Delta t}{r_w^2}, \tag{13-20}$$

where Δt is the shut-in time. The logarithmic approximation of the exponential integral function (with a good accuracy) is valid for small arguments

$$Ei(-x) = \ln x + 0.57722, \qquad x < 0.01. \tag{13-21}$$

From Eqs. (13-20) and (13-21) we obtain the Horner equation

$$T_s\left(r_w,t_s\right) = T_i + M\ln\left(1+\frac{t_c}{\Delta t}\right), \qquad M = \frac{q}{4\pi\lambda}. \tag{13-22}$$

Thus from a semilog plot we can obtain the undisturbed formation temperature and the parameter M. In many cases the dimensionless parameters t_{cD} and t_{sD} are small and Eq. (13-21) cannot be applied. In addition (as was shown by Lachenbruch and Brewer (1959)), the heat source strength at a given depth while drilling may be more realistically considered as a decreasing function of time. It should be also taken into account that drilling records show that the mud is circulating only a certain part of the time required to drilling the well. The evaluation and limitations of the Horner technique are discussed in the literature (Dowdle and Cobb 1975; Drury 1984; Beck and Balling 1988). In Table 13-2 the function $T_D*(t_D) = T_D(t_D)$ (Eq. (13-15)) and the "Exact" solution of Chatas (Lee 1982) are compared.

Thus we can conclude that at small values of t_D (due to short drilling fluid circulation time and low values of thermal diffusivity of formations) the borehole cannot be considered as a linear heat source. Below we present a simple example. Let us assume that: the well radius = 0.1 m, the thermal diffusivity = 0.0040 m²/h, the fluid circulation time = 5 hours, the shut-in time is 2 and 5 hours. Then the value $t_D(t = 1\text{ hr}) = (1\cdot0.0040)/(0.1\cdot0.1) = 0.4$, and in (13-20) the corresponding values of dimensionless time are: $t_D(t = 7\text{ hr}) = 2.8$; $t_D(t = 2\text{ hr}) = 0.8$; $t_D(t = 10\text{ hr}) = 4.0$; $t_D(t = 5\text{ hr}) = 2.0$. From Table 13-2 follows that (13-20) and (13-21) cannot be used to process field data. We consider (13-22) only as an extrapolation formula.

Table 13-2. Comparison of the values of dimensionless wall temperature.

t_D	T_{DCh}	$T_{wD}*$	$T_{DCh}-T_D*$	$R, \%$
0.4	0.5645	0.2161	0.3484	61.71
0.8	0.7387	0.4378	0.3009	40.73
1.0	0.8019	0.5221	0.2798	34.89
1.4	0.9160	0.6582	0.2578	28.14
2	1.0195	0.8117	0.2078	20.38
4	1.2750	1.1285	0.1465	11.49
6	1.4362	1.3210	0.1152	8.02
8	1.5557	1.4598	0.0959	6.17
10	1.6509	1.5683	0.0826	5.01
15	1.8294	1.7669	0.0625	3.42
20	1.9601	1.9086	0.0515	2.63
30	2.1470	2.1093	0.0377	1.76
40	2.2824	2.2521	0.0303	1.33
50	2.3884	2.3630	0.0254	1.06

T_{DCh}–"Exact" solution, $T_{wD}*$– Eq. (13-15), $R = (T_{DCh}-T_D*)/T_{DCh}\cdot100, \%$.

13.4 A New Equation

Using Eq. (13-18) and the principle of superposition for a well as a cylindrical source with a constant heat flow rate $q = q(t_c)$ which operates during the time $t = G\cdot t_c$ and shut-in thereafter (see Chapter 2), we obtain a working formula for processing field data (Kutasov and Eppelbaum 2010):

$$T(r_w,t_s) = T_i + m \ln X,$$ (13-23)

$$X = \frac{1+\left(c - \dfrac{1}{a+\sqrt{Gt_{cD}+t_{sD}}}\right)\sqrt{Gt_{cD}+t_{sD}}}{1+\left(c - \dfrac{1}{a+\sqrt{t_{sD}}}\right)\sqrt{t_{sD}}},$$ (13-24)

$$m = \frac{q}{2\pi\lambda}.$$ (13-25)

The constants a and c were defined earlier (Eq. (13-9)). As can be seen from equation (13-23) the processing of field data (semilog linear log) is similar of that of the Horner method. For this reason we have given the name *"Generalized Horner Method"* (Kutasov and Eppelbaum 2005b) to the procedure just described for determining the static temperature of formations. It is easy to see that for large values of t_{cD} ($G \rightarrow 1$)

and t_{sD} we obtain the well-known Horner equation (Eq. (13-22)). To calculate the ratio X the thermal diffusivity of formations (χ) should be determined with a reasonable accuracy. The effect of variation of this parameter on the accuracy of determining undisturbed formation temperature will be shown below. The value of $\chi = 0.04$ ft²/hr $= 0.0037$ m²/hr was found to be a good estimate for sedimentary rocks (Ramey 1962).

13.4.1 Examples

As will be shown by the following example, it is difficult to determine the accuracy of the Horner method in predicting undisturbed formation temperatures.

Example 1: Basic data used in the example (Schoeppel and Gilarranz 1966) are shown in Table 13-3.

Table 13-3. Shut-in temperatures and data used in Example 1 (Schoeppel and Gilarranz 1966).

No.	t_s, hr	T_s, °F	T_s, °C	t_s/t_c
1	1.0	179.50	81.94	0.333
2	2.0	187.80	86.56	0.667
3	3.0	191.92	88.84	1.000
4	4.0	195.37	90.76	1.333
5	5.0	198.13	92.29	1.667
6	6.0	200.20	93.44	2.000
7	7.0	201.58	94.21	2.333
8	8.0	202.27	94.59	2.667
9	9.0	202.96	94.98	3.000
10	10.0	203.65	95.36	3.333
11	11.0	204.34	95.74	3.667
12	12.0	205.03	96.13	4.000

The example applies to a borehole of 10,000 ft (3,050 m). The geothermal gradient is 1.4°F/100 ft (2.55°C/100 m) and the bottom-hole circulating temperature is determined to be 145°F (62.8°C). The undisturbed formation temperature is 214°F (101.1°C), the well radius is 0.329 ft (0.10 m), and the formation diffusivity is 0.0431 ft²/hr (0.0040 m²/hr). Figure 13-3 shows the computed temperature-time relation (Schoeppel and Gilarranz 1966).

We used Eqs. (13-22) and (13-23) and a linear regression program for the input data processing. The predicted values of T_i are presented in Table 13-4.

The accuracy of the T_i prediction in this example depends on the duration of the shut-in period. For example, if the shut-in period is 3 hours, the $\Delta T_i = 101.11 - 95.70 = 5.41$(°C). At the same time the temperature deviations (R) from the Horner plot are small (Table 13-4).

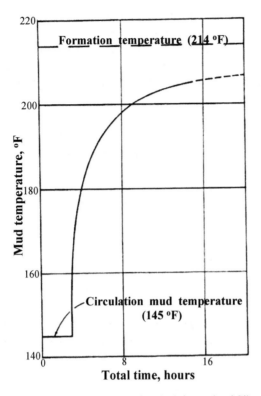

Figure 13-3. Shut-in temperatures for Example 1 as given by Schoeppel and Gilarranz (1966).

Table 13-4. Predicted formation temperature for Example 1.

Combination	(31)			(30)		
	T_f, °C	$-m$, °C	R, %	T_f, °C	$-M$, °C	R, %
1-3	101.04	37.89	0.1	95.70	9.94	0.0
1-4	101.19	38.24	0.1	96.31	10.45	0.2
1-5	101.51	38.99	0.2	96.94	11.01	0.4
1-6	101.77	39.61	0.2	97.45	11.49	0.5
1-7	101.88	39.89	0.2	97.82	11.84	0.6
1-8	101.80	39.68	0.2	98.01	12.03	0.6
1-9	101.68	39.37	0.2	98.13	12.16	0.6
1-10	101.58	39.08	0.2	98.24	12.28	0.5
1-11	101.50	38.86	0.2	98.34	12.38	0.5
1-12	101.46	38.74	0.2	98.45	12.51	0.5

Now let us assume that the thermal diffusivity of the formation is determined with the accuracy of $\pm 20\%$, then for the last combination (12 points) we obtain (after Eq. (13-23)): $\Delta T_i = T_i(\chi = 0.0048 \text{ m}^2/\text{hr}) - T_i(\chi = 0.0040 \text{ m}^2/\text{hr}) = 101.26 - 101.46 = -0.20(^\circ\text{C})$; $\Delta T_i = T_i(\chi = 0.0032 \text{ m}^2/\text{hr}) - T_i(\chi = 0.0040 \text{ m}^2/\text{hr}) = 101.71 - 101.46 = 0.25(^\circ\text{C})$. Thus the effect of variation of thermal diffusivity of formations on the value of T_i can be estimated.

Example 2: This example is from Kelley Hot Springs geothermal reservoir, Moduc County of California. Depth 1035 m (3395 ft) (Roux et al. 1980). The parameter $\chi/r_w^2 = 0.27/\text{hr}$ and $t_c = 12$ hours. The results of temperature measurements and predicted formation temperatures are presented in Table 13-5.

Table 13-5. Predicted formation temperature for Example 2.

t_s, hr	T_s, °C	(31)			(30)		
		T_p, °C	$-m$, °C	R, %	T_p, °C	$-m$, °C	R, %
14.3	83.9	111.32	84.64	0.4	107.26	38.68	0.5
22.3	90.0						
29.3	94.4						

Thus, a new method of determination of formation temperature from bottom-hole temperature logs is developed. It is assumed that the circulating mud temperature is constant. A semi-analytical equation for the transient bore-face temperature during shut-in is presented. At large values of shut-in and mud circulation dimensionless time the suggested equation transforms to the Horner formula (plot).

14

Three Points Method for Estimation of the Formation Temperature

The geothermal measurements (temperature profiles, thermal conductivity of formations) in permafrost regions (Alaska, Northern Canada, Siberia a.o.) can provide a good source of information related to the widely discussed problem of global warming during the last century. Indeed, the anomalies in the temperature profiles (the departure of the temperature profiles from linearity at a constant thermal conductivity of formations) in the permafrost contain a record of change in surface temperature of the past (Osterkamp 1984). In addition, the change in the heat flow density at the permafrost base (frozen-unfrozen interface) is also an indicator of the climate change in the past (Melnikov et al. 1973). When interpreted with the heat conduction theory, these sources can provide important information of patterns of contemporary climate change. For example, precision measurements in oil wells in the Alaskan Arctic indicate a widespread warming (2–4°C) at the permafrost surface during the 20th century (Lachenbruch et al. 1988).

In permafrost regions, due to thawing of the surroundings of the wellbore formations during drilling, data representative of undisturbed geothermal temperature can be obtained only by repeated observations over a long period of time (Eppelbaum et al. 2014). The drilling process greatly alters the temperature field of formations surrounding the wellbore. The temperature change is affected by the duration of fluid circulation (depth penetration, hole cleaning, cementing), the duration of shut-in periods (tripping of drill pipe, running of casing, logging a.o.), the temperature difference between the formation and drilling mud, the well radius, the thermal properties of formations, and the drilling technology used.

The results of field and analytical investigations have shown that in many cases the effective temperature (T_w) of the circulating fluid (mud) at a given depth can be assumed constant during drilling or production (Lachenbruch and Brewer 1959; Ramey 1962; Edwardson et al. 1962; Jaeger 1961; Kutasov et al. 1966; Raymond 1969). Here we should note that even for a continuous mud circulation process the wellbore temperature is dependent on the current well depth and other factors. The term "effective fluid temperature" is used to describe the temperature disturbance of formations while drilling. In their classical paper, Lachenbruch and Brewer (1959) have shown that the wellbore shut-in temperature mainly depends on the amount of thermal energy transferred to (or from) formations. The thermal effect of drilling operations on the temperature field of formations can be also approximated by a constant cylindrical source with a contact thermal resistance (Wilhelm et al. 1995). The mud circulation time and the total well drilling time ratio decrease with the increase of the well depth. For this reason, for deep wells both models (constant temperature, constant cylindrical/linear source) cannot be used to describe the thermal disturbance of formations and the temperature recovery during the shut-in period. In addition, in an actual drilling process a number of time dependent variables influence downhole temperatures. The composition of annular materials (steel, cement, fluids), the drilling history (vertical depth versus time), the duration of short shut-in periods, fluid flow history, radial and vertical heat conduction in formations, the change of geothermal gradient with depth, and other factors should be accounted for and their effects on the wellbore temperatures while drilling should be determined (Kutasov 1976, 1999; Eppelbaum et al. 1996, 2014). It is clear that only transient computer models can be used to calculate temperatures in the wellbore and surrounding formations as functions of depth and time. However, in many cases it is difficult to compare the results of a computer simulation with actual temperatures measured in wells during circulation and/or shut-in periods. Indeed, the data accompanying field-collected temperature measurements often are incomplete. For example, it is quite common not to have the mud composition and even the mud type. Also, in some cases, very little is known about the type of formations penetrated by the well.

The objective of this chapter is to suggest a new approach in utilization temperature logs in deep wells and present working formulas for determining the undisturbed formation temperature. Below we will introduce a new approach in predicting the undisturbed formation temperatures (and geothermal gradients) from shut-in temperature logs in deep wells.

The main features of the suggested method are:

1. In the permafrost section of the studied well the starting point in the well thermal recovery is moved from the end of well completion to the moment of time when the refreezing of formations was completed. It is taken into account that the refreezing of thawed formations occurs in some temperature interval.
2. Below the permafrost base the starting point in the well thermal recovery is moved from the end of well completion to the moment of time when the first shut-in temperature log was taken.

Thus the application of the proposed method of predicting the undisturbed formation temperature does not depend:

(a) on the well drilling history (vertical depth versus time, stops in mud circulation), drilling technology used (properties of drilling fluids, penetration rate, bit size, casings, cementing techniques),
(b) on the duration of the complete refreezing formations thawed during drilling.

14.1 Shut-in Temperatures–Permafrost Zone

Let us assume that at the moment of time $t = t_{ep}$ the phase transitions (water-ice) in formations at a selected depth $z < z_{pf}$ (depth to base of ice-bounded permafrost) are complete, i.e., the thermally disturb formation has frozen. In this case at $t > t_{ep}$ the cooling process is similar to that of temperature recovery in sections of the well below the permafrost base. It is well known (Tsytovich 1975) that the freezing of the water occurs in some temperature interval below 0°C (Figure 14-1).

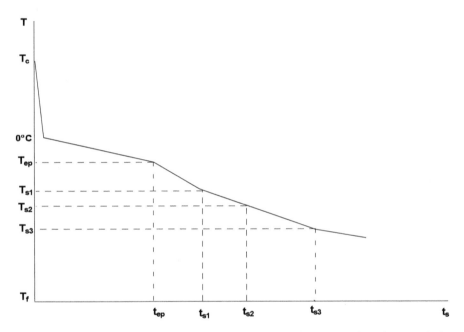

Figure 14-1. Shut-in temperatures at a given depth (above the permafrost base)—schematic curve. t_s is the shut-in time, and T_s is the shut-in temperature.

In practice, however, the moment of time $t = t_{ep}$ cannot be determined. This can be done only by conducting long-term repetitive temperature observations in deep wells.

14.2 Working Equation

Let us assume that three shut-in temperatures T_{s1}, T_{s2}, and T_{s3} are measured at a given depth (Figure 14-1). We can consider that the period of time $t_c^* = t_c + t_{s1}$ as a new "thermal disturbance" period. Then the "shut-in times" are

$$t_{s1}^* = t_{s2} - t_{s1}, \quad t_{s2}^* = t_{s3} - t_{s1}. \tag{14-1}$$

Now dimensionless temperature distribution at $t = t_{s1}$

$$
\begin{cases}
T_{cD}^*\left(r_D, t_{xD}\right) = 1, & 0 \le r_D \le 1, \\
T_{cD}^*\left(r_D, t_{xD}\right) = 1 - \dfrac{\ln r_D}{\ln R_x}, & 1 \le r_D \le R_x, \\
T_{cD}^*\left(r_D, t_{xD}\right) = 0, & r_D > R_x.
\end{cases} \tag{14-2}
$$

$$
R_x = 1 + 2.184\sqrt{t_{xD}}, \qquad t_{xD} = \frac{a_f t_c^*}{r_{wx}^2}, \qquad R_x = \frac{r_{ix}}{r_{wx}}, \tag{14-3}
$$

$$
T_{cD}^*\left(r_D, t_{xD}\right) = \frac{T_c\left(r, t\right) - T_f}{T_{s1} - T_f}, \qquad r_D = \frac{r}{r_{wx}}, \tag{14-4}
$$

where r_{wx} is the radius of a cylindrical source with a constant wall temperature (T_{s1}) during the thermal disturbance period (t_c^*), a_f is the thermal diffusivity of frozen formations, t_c is the time of "thermal disturbance" (at a given depth) during drilling, T_f is the undisturbed temperature of formations, and r_{ix} is the radius of thermal influence.

It is a reasonable assumption that the value of t_c (the "disturbance" period) is a linear function of the depth (z):

$$
t_c = t_{tot}\left(1 - \frac{z}{H_t}\right), \tag{14-5}
$$

where z is the depth, t_{tot} is the total drilling time, and H_t is the total well depth.

For the initial radial temperature distributions (Eq. (14-2)) the dimensionless shut-in wellbore temperature (at $t_s > t_{s1}$) was presented earlier (Kritikos and Kutasov 1988; Kutasov 1999):

$$
T_{sD}^* = \frac{T\left(0, t_s^*\right) - T_f}{T_{s1} - T_f} = 1 - \frac{Ei\left(-p^* R_x^2\right) - Ei\left(-p^*\right)}{2 \ln R_x}, \tag{14-6}
$$

$$
p^* = \frac{1}{4n^* t_{xD}}, \qquad n^* = \frac{t_s^*}{t_c^*}, \tag{14-7}
$$

where Ei is the exponential integral.

By using measurements T_{s2}, T_{s3} and Eq. (14-6), we can eliminate the formation temperature T_f. After simple transformations we obtain

$$
\gamma = \frac{T_{s2} - T_{s1}}{T_{s3} - T_{s1}} = \frac{Ei\left(-p_1^* R_x^2\right) - Ei\left(-p_1^*\right)}{Ei\left(-p_2^* R_x^2\right) - Ei\left(-p_2^*\right)}, \tag{14-8}
$$

where

$$
p_1^* = \frac{1}{4n_1^* t_{xD}}, \qquad n_1^* = \frac{t_{s1}^*}{t_c^*}, \qquad p_2^* = \frac{1}{4n_2^* t_{xD}}, \qquad n_2^* = \frac{t_{s2}^*}{t_c^*}. \tag{14-9}
$$

Substituting the value of R_x (Eq. (14-3)) into Eq. (14-8) we can obtain a formula for calculating the dimensionless disturbance time, t_{xD}. After this is possible to determine values of T_f, R_x, and $A = a_f/(r_{wx})^2$.

Although Eq. (14-8) is based on an analytical solution (Eq. (14-6)), we should mention several limitations in application of the suggested method. Firstly, the temperature ratio γ (Eq. (14-8)) should be determined with high accuracy. This means that high accuracy of temperature measurements (T_{s1}, T_{s2}, T_{s3}) is needed. The temperature differences $T_{s2} - T_{s1}$ and $T_{s3} - T_{s1}$ should be significantly larger than the absolute accuracy of temperature measurements. For this reasons the proposed method cannot be used for depths where the undisturbed formation temperature is close to 0°C. Secondly, the lithological profile of the permafrost section of the well should be known. This will allow to find the temperature interval of thawed formation refreezing, and to select the "initial" temperature log (T_{s1}, Figure 14-1). To determine the depth of the 0°C isotherm (the position of the permafrost base) we recommend using the "two point" method (Kutasov 1988). The "two point" method of predicting the permafrost thickness is based on determining the geothermal gradient in a uniform layer below the permafrost zone. Therefore, a lithological profile for this section of the well must be available. Only two shut-in temperature measurements for two depths are needed to determine the geothermal gradient. The position of the permafrost base is predicted by the extrapolation of the undisturbed formation temperature-depth curve to 0°C. To use a computer program Eq. (14-8) should be rewritten as

$$\gamma - \frac{Ei\left(-p_1^* R_x^2\right) - Ei\left(-p_1^*\right)}{Ei\left(-p_2^* R_x^2\right) - Ei\left(-p_2^*\right)} = \varepsilon, \tag{14-8)*}$$

where ε is a small value and depends on the accuracy of the γ ratio. The Newton method was used (Grossman 1977) for solving Eq. (14-8)*. In this method a solution of an equation is sought by defining a sequence of numbers which become successively closer and closer to the solution. The conditions, which guarantee that Newton method in our case will work and provide a unique solution, are satisfied (Grossman 1977).

In the subroutine which utilizes the Newton method the following parameters were used:

(a) the starting value of t_{xD} is 0.01; (b) the time increment is 2.0; (c) the absolute accuracy of the ratio of y is $\varepsilon = 0.00001$. To speed up calculations we prepared a computer program "PERMTEMP" (Kutasov 1999, pp. 318–320). This program was utilized to process field data for all five wells.

14.3 Unfrozen Well Section—the Initial Temperature Distribution

In theory the drilling process affects the temperature field of formations at very long radial distances. There is however, a practical limit to the distance—the radius of thermal influence (r_{in}), where for a given circulation period ($t = t_c$) the temperature $T(r_{in}, t_c)$ is "practically" equal to the geothermal temperature T_f. To avoid uncertainty, however, it is essential that the parameter r_{in} must not to be dependent on the temperature difference $T(r_{in}, t_c) - T_f$. For this reason we used the thermal balance method to calculate the radius of thermal influence. The results of modeling, experimental works and field observations have shown that the temperature distribution around the wellbore during drilling can be approximated by the following relation (Kutasov 1968, 1999):

$$\frac{T(r,t) - T_f}{T_w - T_f} = 1 - \frac{\ln r/r_w}{\ln r_{in}/r_w}, \qquad r_w \leq r \leq r_{in}, \tag{14-10}$$

where T_w is the effective temperature of the drilling fluid (at a given depth). Introducing the dimensionless values of circulation time, radial distance, radius of thermal influence, and temperature

$$t_D = \frac{a_f t_c}{r_w^2}, \qquad r_D = \frac{r}{r_w}, \qquad R_{in} = \frac{r_{in}}{r_w}, \qquad T_D(r_D, t_D) = \frac{T(r,t) - T_f}{T_w - T_f}, \tag{14-11}$$

we obtain

$$T_D(r_D, t_D) = 1 - \frac{\ln r_D}{\ln R_{in}}, \qquad 1 \leq r_D \leq R_{in}, \tag{14-12}$$

$$R_{in} = \frac{r_{in}}{r_w} = 1 + D_0\sqrt{t_D}, \qquad D_0 = 2.184, \qquad 5 \leq t_D \leq 10^4, \tag{14-13}$$

Where a_f is the thermal diffusivity of the formation (Kutasov 1968, 1999). The radius of thermal influence r_{in} in Eq. (14-13) was determined from the thermal balance condition (Figure 14-2).

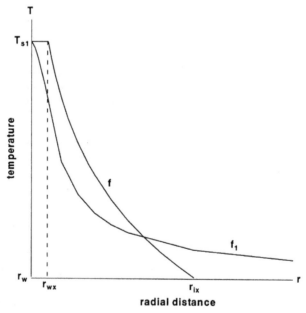

Figure 14-2. Actual (curve f_1) and assumed (curve f) radial temperature distributions by $t_s = t_{s1}$ at a given depth (below the permafrost base)—schematic curves.

Again, the value of t_c (the "disturbance" period) can be estimated from Eq. (14-5). Let us also now assume that three shut-in temperature T_{s1}, T_{s2}, and T_{s3} are measured at a given depth. We can consider that the period of time $t^*_c = t_c + t_{s1}$ as a new "thermal disturbance" period and r_{wx} as an "effective" well radius:

$$\int_{r_w}^{\infty} r f_1\left(t_{s1}, r\right) dr = \int_{r_{wx}}^{r_{ix}} r f\left(t_{s1}, r\right) dr. \tag{14-14}$$

Now the dimensionless radial temperature distribution at $t_s = t_{s1}$ is also expressed by Eq. (14-2). Hence, for deep depths (below the permafrost base) Eq. (14-6) can be used for calculations of the undisturbed formations temperatures and geothermal gradients.

At the moment of shut-in time t_{s1} the actual radial temperature distribution is f_1 (Figure 14-2).

14.4 Field Cases

Extensive temperature measurements in the Northern Canada and Alaska were conducted by the Geothermal Service of Canada (Taylor et al. 1977; Judge et al. 1979, 1981; Taylor et al. 1982) and U.S. Geological Survey (Lachenbruch and Brewer 1959; Boreholes 1998). We selected long term temperature surveys in five wells to verify proposed Eq. (14-6).

Well 3, Alaska, South Barrow (Lachenbruch and Brewer 1959)

Temperature measurements to a depth of 595 feet were made during a period of six years after drilling. The well was drilled for 63 days to a total depth of 2,900 ft. The predicted by Lachenbruch and Brewer (1959) equilibrium formation temperatures are: −8.835°C, −7.830°C, and −6.735°C for depths 355, 475, 595 ft respectively. On the average of 0.2°C (for 31 runs) our results differ from these values (Table 14-1).

Our model does not take into account the effect of the geothermal gradient on the restoration of the natural temperature field of formations. This may contribute to difference of 0.2°C. We should note that at calculations of T_f (Table 14-1) we used temperature measurements with short shut-in times. For the depth of 595 ft (Figure 14-3) the average value of T_f is 6.568°C (9 runs). The completion of freezing occurred at temperature of about −0.6°C and duration of the complete freezeback is approximately 20 days (Figure 14-3).

For processing field temperature data Lachenbruch and Brewer (1959) used the following empirical formula

$$T_w = C \ln\left(1 + \frac{t_c}{t_s}\right) + B, \tag{14-15}$$

where C and B are empirical coefficients determined from field measurements.

The last formula and Horner equation are identical and at $t_s \to \infty$ and $T_w \to T_f$.

Table 14-1. Observed shut-in temperatures T_{s1}, T_{s2} and T_{s3} (Lachenbruch and Brewer 1959) and calculated formation temperature T_f. Alaska, South Barrow Well 3.

Run No.	T_{s1}, °C	T_{s2}, °C	T_{s3}, °C	$t_{s1}{}^*$, day	$t_{s2}{}^*$, day	$t_{s3}{}^*$, day	t_{xD}	A, 1/day	R_x	T_f °C
				Depth 595 ft. $t_c = 50$ days						
1	−2.163	−3.345	−4.829	80	87	117	13.8	0.1721	9.10	−6.541
2	−2.163	−3.345	−5.155	80	87	133	13.7	0.1712	9.08	−6.552
3	−2.163	−3.345	−5.542	80	87	167	13.8	0.1726	9.12	−6.536
4	−2.163	−3.345	−5.764	80	87	197	13.6	0.1701	9.06	−6.567
5	−2.163	−3.345	−6.191	80	87	338	13.3	0.1667	8.98	−6.609
6	−3.345	−4.829	−5.155	87	117	133	6.59	0.0758	6.61	−6.574
7	−3.345	−4.829	−5.542	87	117	167	6.89	0.0792	6.73	−6.533
8	−3.345	−4.829	−5.764	87	117	197	6.55	0.0752	6.59	−6.580
9	−3.345	−4.829	−6.191	87	117	338	6.27	0.0721	6.47	−6.621
				Depth 475 ft. $t_c = 51$ days						
10	−2.741	−3.780	−4.452	74	81	88	8.66	0.1164	7.43	−7.648
11	−2.741	−3.780	−5.751	74	81	118	8.65	0.1161	7.42	−7.654
12	−2.741	−3.780	−6.079	74	81	134	8.69	0.1167	7.44	−7.638
13	−2.741	−3.780	−6.517	74	81	168	8.61	0.1156	7.41	−7.669
14	−2.741	−3.780	−6.752	74	81	199	8.59	0.1153	7.40	−7.677
15	−2.741	−3.780	−7.219	74	81	339	8.49	0.1140	7.36	−7.712
16	−3.780	−4.452	−5.751	81	88	118	8.05	0.0993	7.20	−7.376
17	−3.780	−4.452	−6.079	81	88	134	7.87	0.0970	7.13	−7.437
18	−3.780	−4.452	−6.517	81	88	168	7.57	0.0933	7.01	−7.542
19	−3.780	−4.452	−6.752	81	88	199	7.46	0.0920	6.97	−7.582
20	−3.780	−4.452	−7.219	81	88	339	7.25	0.0894	6.88	−7.666
				Depth 355 ft. $t_c = 56$ days						
21	−4.550	−5.565	−6.122	79	86	93	9.92	0.1257	7.88	−8.845
22	−4.550	−5.565	−7.141	79	86	123	10.5	0.1335	8.09	−8.702
23	−4.550	−5.565	−7.380	79	86	139	10.8	0.1361	8.16	−8.658
24	−4.550	−5.565	−7.706	79	86	173	10.9	0.1378	8.20	−8.632
25	−4.550	−5.565	−7.890	79	86	204	10.9	0.1375	8.20	−8.636
26	−4.550	−5.565	−8.289	79	86	344	10.6	0.1341	8.11	−8.692
27	−5.565	−6.122	−7.141	86	93	123	9.06	0.1050	7.57	−8.435
28	−5.565	−6.122	−7.380	86	93	139	8.93	0.1036	7.53	−8.463
29	−5.565	−6.122	−7.706	86	93	173	8.73	0.1012	7.45	−8.509
30	−5.565	−6.122	−7.890	86	93	204	8.58	0.0995	7.40	−8.544
31	−5.565	−6.122	−8.289	86	93	344	8.18	0.0948	7.24	−8.649

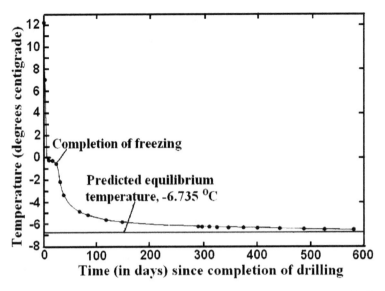

Figure 14-3. Cooling of South Barrow (Alaska) well 3, 595-foot depth-temperature curve (Lachenbruch and Brewer 1959).

Well No. 175 (Judge et al. 1981). The location and input data of this well are presented in Table 14-2. Figure 14-4 is a graph of the 7 temperature surveys in this well. The field data have shown that the duration of refreezing of formation thawed out during drilling significantly depends on its natural (undisturbed) temperature (or depth). For low temperature permafrost the refreezing period is relatively short (Figure 14-4). The restoration of the temperature regime was accompanied by a nearly isothermal interval (Figure 14-4).

Table 14-2. Input data and location of the well No. 175 (Judge et al. 1981).

Site name	Gemini E-10
Latitude	79° 59'24" N
Longitude	84° 04'12" W
Hole depth, m	3845
Well spudded	14-10-72
Drilling time, days	145

Therefore, if the shut-in time is insufficient, one may incorrectly attribute the zero temperature gradient intervals to some geological-geographical factors. It was found from long-term temperature surveys that only after 28 ($z = 100$ m) and 17 years ($z = 650$ m) of shut-in periods the undisturbed formation temperature can be measured with an accuracy of 0.1°C (Judge et al. 1981). Comparison of predicted undisturbed formation temperatures (Table 14-3) and those (T_{EQ}) determined from 7 long-term temperature logs (Judge et al. 1981; p. 131) are in a fair agreement.

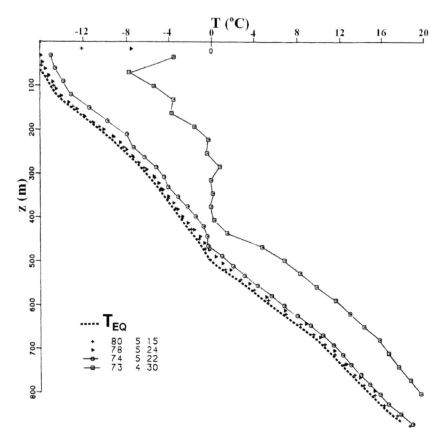

Figure 14-4. Temperature profiles in the well No. 175, T_{EQ} is the predicted temperature versus depth curve (Judge et al. 1981).

Table 14-3. Observed shut-in temperatures T_{s1}, T_{s2}, and T_{s3} (Judge et al. 1981) and calculated formation temperature T_f and T_{EQ}. Well No. 175, $t_{s1} = 53$, $t_{s1} = 440$, and $t_{s1} = 795$ days.

z, m	t_{xD}	A, 1/day	R_x	T_{s1}, °C	T_{s2}, °C	T_{s3}, °C	T_f, °C	T_{EQ}, °C
101.5	15.47	0.07968	9.59	−5.44	−13.57	−14.13	−14.86	−15.26
132.3	20.32	0.10525	10.84	−3.58	−12.44	−13.00	−13.72	−14.25
162.8	5.17	0.02695	5.97	−3.74	−10.74	−11.44	−12.38	−12.62
559.0	16.68	0.09431	9.92	9.79	4.47	4.14	3.72	3.32
589.5	58.25	0.33142	17.67	11.59	6.20	5.96	5.66	5.05
620.3	32.94	0.18864	13.53	12.96	7.84	7.58	7.25	6.78
650.1	9.95	0.05735	7.89	14.30	9.43	9.08	8.63	8.44

Interpolation of temperature data was used to determine the values of T_{s2}, T_{s3}, and T_{EQ} (Table 14-3). The authors used the empirical Lachenbruch-Brewer formula (Horner method) to estimate the undisturbed formation temperature (T_{EQ}). We should also note that it is difficult to determine the accuracy of the Horner method in predicting undisturbed formation temperature. The accuracy of T_{EQ} prediction, as was shown in one example, depends on the duration of the shut-in period. At the same time the temperature deviations from the Hornet plot were small (Kutasov 1999, p. 259). Our model does not take into account the effect of the geothermal gradient on the restoration of the natural temperature field of formations. This may have some effect on the predicted values of T_f in regions with high geothermal gradients. In Figure 14-5 the predicted and measured shut-in temperatures at two depths are compared.

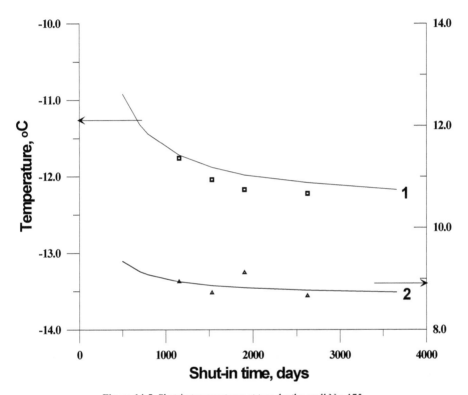

Figure 14-5. Shut-in temperatures at two depths, well No. 175.

Curve 1—162.8 m; Curve 2—650.1 m, t_{s1} = 53, t_{s2} = 440, and t_{s3} = 795 days. Solid curves—calculated values, points—observed temperatures (interpolated).

It can be seen that Eq. (14-6) gives a sufficiently accurate description of the process by which temperature equilibrium comes about in the borehole (Table 14-3, Figure 14-5).

The temperature gradient is a differential quantity hence the process by which the geothermal (undisturbed) gradient (Γ) is restored is distinct from the temperature recovery process. Eq. (14-6) can be used to answer the question: how accurately is the value of Γ determined from temperature measurements taken a short time after cessation of drilling.

From analysis of Figure 14-6 follows that in the 589.5–650.1 m interval a shut-in of 800 days is required to determine the value of Γ with the accuracy of 5%.

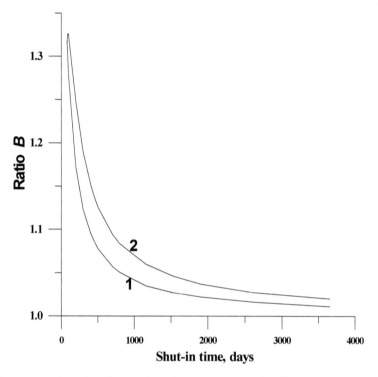

Figure 14-6. The geothermal gradient Γ and transient temperature gradient (G) ratio (B) for two sections, well No. 175. Curve 1 – 101.5–162.8 m, $G(t_s = \infty) = \Gamma = 0.04049$ m/K, Curve 2—589.5–650.1 m, $G(t_s = \infty) = \Gamma = 0.04901$ m/K.

West Fish Creek well No. 1 (Boreholes 1998). The information about this well and the results of temperature surveys are presented in Tables 14-4 and 14-5.

Table 14-4. Input data and location of three wells, Alaska (Boreholes 1998).

Site code	FCK	PBF	DRP
Site name	West Fish Creek No. 1	Put River N-1	Drew Point No. 1
Latitude	70° 19'35.99" N	70° 19'07" N	70° 52' 47.14" N
Longitude	152° 03' 38.03" W	148° 54'35" W	153°53' 59.93" W
Surface elevation, m	27	8	5
Casing diameter, cm	34	51	34
Hole depth, m	735	763	640
Date of drill start	14-02-77	02-09-70	13-01-78
Drilling time, days	73	44	60

Table 14-5. Results of temperature surveys (°C) in the West Fish Creek well No. 1, Alaska. Boreholes (1998).

z, m	Shut-in time, days					
	227	506	867	1232	1576	2657
30.48	−7.018	−7.815	−8.259	−8.411	−8.534	−8.831
60.96	−5.721	−7.134	−7.676	−7.849	−7.978	−8.094
91.44	−5.055	−6.285	−6.901	−7.065	−7.212	−7.409
112.78	−4.818	−5.584	−6.155	−6.285	−6.406	−6.625
146.30	−3.624	−4.196	−4.719	−4.795	−4.946	−5.103
179.83	−1.617	−2.391	−2.977	−3.053	−3.241	−3.423
210.31	−0.104	−0.924	−1.511	−1.587	-	−1.961
240.79	0.955	0.248	−0.235	−0.291	-	−0.666
271.27	2.161	1.437	0.984	0.892	0.718	0.539
301.75	3.220	2.523	2.071	2.015	1.829	1.662
332.23	4.315	3.643	3.221	3.147	2.989	2.800
368.81	5.861	4.972	4.563	4.543	4.348	4.182
399.29	6.818	6.165	5.743	5.702	5.503	5.335
432.82	7.978	7.346	6.954	6.955	6.731	6.563
478.54	9.656	9.020	8.587	8.597	8.333	8.157
512.07	10.647	10.033	9.650	10.000	9.456	9.284
542.55	12.014	11.439	11.067	11.099	10.864	10.658
573.03	13.023	12.433	12.111	12.136	11.926	11.739
606.55	14.191	13.618	13.308	13.362	13.130	12.934
640.08	15.225	14.838	14.552	14.606	14.329	14.176
670.56	16.357	15.843	15.572	15.578	15.362	15.238
704.09	17.935	17.431	17.137	17.188	16.931	16.705

The computer program "PERMTEMP" (FORTRAN) was used to calculate the values of undisturbed formation temperatures for a frozen/unfrozen formations sections of the well (Table 14-6).

Let us estimate the values of r_{wx} and r_{ix} for the depth of 112.78 m. From Table 14-6 we find that $A = a_f(r_{wx})^2 = 0.002250$ 1/day. Let assume that the thermal diffusivity of the frozen formation is $a_f = 8.333 \cdot 10^{-7}$ m²/s = 0.0030 m²/hr, then

$$r_{wx} = \sqrt{24 \cdot 0.0030 / 0.002250} = 5.657 \text{ (m)},$$

$$r_{ix} = 5.657 + 2.184 \cdot 5.657 \cdot \sqrt{0.650} = 15.62 \text{ (m)}.$$

Table 14-6. Predicted formation temperatures, well West Fish Creek well No. 1, Alaska; $t_{s1} = 227$, $t_{s2} = 506$, and $t_{s3} = 867$ days.

z, m	t_{xD}	A, 1/day	R_x	T_p °C
30.48	1.011	0.003405	3.196	−8.898
60.96	1.826	0.006212	3.951	−8.374
91.44	1.155	0.003970	3.347	−7.754
112.78	0.650	0.002250	2.761	−7.063
146.30	0.497	0.001740	2.539	−5.623
179.83	0.618	0.002191	2.717	−3.910
368.81	1.144	0.004342	3.335	4.022
399.29	0.685	0.002632	2.808	5.120
432.82	0.712	0.002771	2.843	6.385
478.54	0.614	0.002431	2.711	7.939
512.07	0.678	0.002720	2.798	9.097
542.55	0.635	0.002579	2.740	10.524
573.03	0.790	0.003248	2.941	11.668
606.55	0.784	0.003271	2.934	12.884
640.08	0.507	0.002144	2.555	14.116
670.56	0.784	0.003359	2.934	15.206
704.09	0.667	0.002898	2.783	16.728

Similarly, for the depth of 704.09 m (here a_f is the thermal diffusivity of the formation) we obtain:

$$r_{wx} = \sqrt{24 \cdot 0.0030 / 0.002898} = 4.984 \text{ (m)}$$

and

$$r_{ix} = 4.984 + 2.184 \cdot 4.984 \cdot \sqrt{0.667} = 13.87 \text{ (m)}.$$

From Table 14-7 one can see the accuracy of shut-in temperature predictions. Let us determine the average squared temperature deviation:

$$\Delta T = \sqrt{\frac{\sum\limits_{n=1}^{N} (T_n^* - T_n)^2}{N}},$$

where T^* is the observed temperature, T is the calculated temperature, and N is the number of readings. In our case $N = 23$ (Table 14-7) and $\Delta T = 0.099$°C.

Table 14-7. Observed (T^*) and calculated (T) temperatures (°C) at three shut-in times, the West Fish Creek No. 1 well, Alaska.

z, m	Shut-in time, days					
	1232		1576		2657	
	T	T^*	T	T^*	T	T^*
30.48	−8.449	−8.411	−8.547	−8.534	−8.691	−8.831
60.96	−7.891	−7.849	−8.000	−7.978	−8.156	−8.094
91.44	−7.158	−7.065	−7.290	−7.212	−7.481	−7.409
112.78	−6.415	−6.285	−6.554	−6.406	−6.759	−6.625
399.29	5.559	5.702	5.464	5.503	5.324	5.335
512.07	9.486	-	9.400	9.456	9.276	9.284
606.55	13.180	13.362	13.114	13.130	13.019	12.934
704.09	17.014	17.188	16.950	16.931	16.859	16.705

The Put River N-1 well. In Figure 14-7 we present some results of temperature surveys in this well. It can be seem from Table 14-8 that Eq. (14-6) approximates the transient shut-in temperatures with a good accuracy. Indeed, the calculated values of ΔT (the average squared temperature deviation) are: 0.070, 0.032, and 0.068°C for depths 45.72, 60.96, and 91.44 m, respectively.

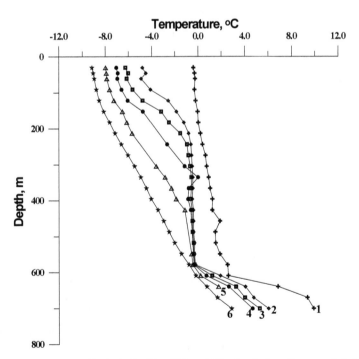

Figure 14-7. Temperature profiles in the Put River N-1 well. The shut-in times for curves 1, 2, 3, 4, 5, and 6 are 5, 34, 48, 66, 117, and 1071 days respectively (Boreholes 1998).

Table 14-8. Observed (T^*) and calculated (T) shut-in temperatures (°C) at three depths of the Put River N-1 well, Alaska; $t_{s1} = 34$, $t_{s2} = 48$, and $t_{s3} = 66$ days.

t_s, days	45.72 m		60.96 m		91.44 m	
	T	T^*	T	T^*	T	T^*
91	−7.546	−7.511	−7.525	−7.497	−7.258	−7.227
117	−7.921	−7.900	−7.867	−7.860	−7.651	−7.620
163	−8.294	−8.428	−8.207	−8.263	−8.040	−7.965
1071	−9.098	−9.052	−8.946	−8.957	−8.875	−8.771

The Drew Point No. 1 well. In Tables 14-9 and 14-10 are presented the results of temperature surveys and predicted values of undisturbed formation temperatures.

Table 14-9. Results of temperature surveys (°C) in the Drew Point No. 1 well, Boreholes 1998.

z, m	shut-in time, days					
	186	547	907	1259	2000	2339
50.29	−7.373	−8.519	−8.726	−8.736	−8.937	−8.953
70.10	−6.925	−8.142	−8.355	−8.292	−8.552	−8.574
100.58	−5.892	−7.283	−7.470	−7.607	−7.709	−7.774
120.40	−5.309	−6.638	−6.852	−6.953	−7.094	−7.140
150.88	−4.466	−5.690	−5.856	−5.961	−6.096	−6.148
170.69	−3.951	−5.076	−5.233	−5.335	−5.481	−5.518
199.64	−3.247	−4.199	−4.342	−4.441	−4.571	−4.619
219.46	−2.698	−3.639	−3.760	−3.860	−3.977	−4.032

Table 14-10. Predicted formation temperatures, well Drew Point No. 1, Alaska.
$t_{s1} = 186$, $t_{s2} = 547$, and $t_{s3} = 907$ days.

z, m	t_{xD}	A, 1/day	R_x	T_f, °C
50.29	2.727	0.011302	4.607	−9.009
70.10	2.874	0.012002	4.702	−8.644
100.58	5.132	0.021692	5.947	−7.715
120.40	3.315	0.014124	4.976	−7.138
150.88	4.813	0.020758	5.791	−6.073
170.69	4.418	0.019210	5.591	−5.439
199.64	3.636	0.015997	5.164	−4.531
219.46	5.165	0.022913	5.964	−3.916

One can see from Figure 14-8 that the observed and calculated values of transient shut-in temperatures are in fair agreement.

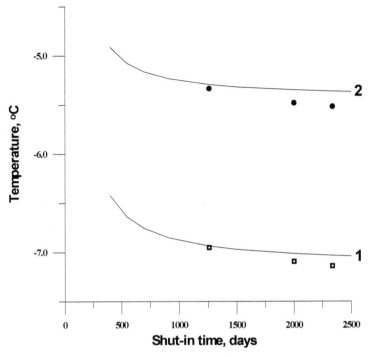

Figure 14-8. Shut-in temperatures at two depths, well Drew Point No. 1, Alaska.

Curves 1 – 120.40 m; 2 – 170.69 m, t_{s1} = 186, t_{s2} = 547, and t_{s3} = 907 days. Solid curves – calculated values, points – observed temperatures.

For the depths 120.40 and 170.69 m the calculated values of ΔT are 0.078 and 0.119°C, respectively.

14.5 Laboratory Experiment

An electrical heater was used by Dr. V. Devyatkin to simulate the drilling process (Kutasov 1976). The heater operated for 91 hours in a model of a well placed in the underground (depth 15 m) laboratory of Permafrost Institute (Yakutsk, Russia). High temperature-sensitive thermistors were used to monitor the shut-in temperatures in the "borehole" (Table 14-11) and surrounding formations (sands).

It was determined that only at –1°C the freezing of the water is practically completed. The last three runs (Table 14-11) show that, when a temperature measurement (–0.37°C) is taken while the refreezing process continues, the value of T_f cannot be determined with a sufficient accuracy.

On the basis of aforementioned we can conclude that a new approach in predicting the undisturbed formation temperature from temperature logs in deep wells is suggested. The proposed method of predicting the undisturbed formation temperature does not depend on the well drilling history, drilling technology used, and on the duration of the complete refreezing formations thawed during drilling. To

Table 14-11. Experimental shut-in temperatures T_{s1}, T_{s2}, T_{s3} (Kutasov 1999) and calculated formation temperature T_f.

No.	T_{s1}, °C	T_{s2}, °C	T_{s3}, °C	t_{s1}, hr	t_{s2}, hr	t_{s3}, hr	t_{xD}	A, 1/hr	R_x	T_f
1	−1.35	−2.21	−2.82	54	57	66	125.0	0.864	25.4	−4.35
2	−1.35	−2.21	−3.03	54	57	72	117.0	0.805	24.6	−4.42
3	−1.35	−2.21	−3.09	54	57	78	140.0	0.969	26.9	−4.24
4	−1.35	−2.21	−3.33	54	57	96	141.0	0.974	27.0	−4.23
5	−1.35	−2.21	−3.49	54	57	120	149.0	1.028	27.7	−4.18
6	−1.35	−2.21	−3.60	54	57	144	152.0	1.045	27.9	−4.17
7	−2.21	−2.82	−3.33	57	66	96	33.7	0.228	13.7	−4.15
8	−2.21	−3.03	−3.49	57	72	120	49.6	0.335	16.4	−4.06
9	−2.21	−3.09	−3.60	57	78	144	30.3	0.205	13.0	−4.15
10	−2.52	−2.82	−3.33	60	66	96	21.4	0.142	11.1	−4.09
11	−2.52	−3.03	−3.49	60	72	120	26.7	0.177	12.3	−4.05
12	−2.52	−3.09	−3.60	60	78	144	16.1	0.107	9.77	−4.16
13	−0.37	−1.35	−2.21	51	54	57	23.2	0.164	11.5	−9.11
14	−0.37	−1.35	−2.52	51	54	60	30.9	0.217	13.1	−7.27
15	−0.37	−1.35	−2.71	51	54	63	36.0	0.253	14.1	−6.56

process field data a generalized formula (for the well sections below and above the permafrost base) is presented. Temperature logs conducted in five wells were used to verify this method. It was found that the predicted and observed transient shut-in temperatures are in a good agreement.

15

Two Logs Method

The mathematical model of the Two Logs Method (TLM) is based on the assumption that in deep wells the effective temperature of drilling mud at a given depth can be assumed to be constant during the drilling process (Kutasov 1968; Kritikos and Kutasov 1988; Kutasov 1999). As was shown earlier (Kutasov 1999) for moderate and large values of the dimensionless circulation time ($t_D > 5$) the temperature distribution function $T_{cD}(r_D, t_D)$ in the vicinity of the well can be described by a simple formula

$$T_{cD}(r_D,t_D) = \frac{T(r_D,t_D) - T_f}{T_w - T_f} = 1 - \frac{\ln r_D}{\ln R_{in}}, \qquad 1 \le r_D \le R_{in}, \qquad (15\text{-}1)$$

$$t_D = \frac{\chi t_c}{r_w^2}, \qquad r_D = \frac{r}{r_w}, \qquad R_{in} = \frac{r_{in}}{r_w},$$

$$R_{in} = 1 + D_o \sqrt{t_D},$$
$$D_o = 2.184, \qquad 5 \le t_D < 10^4,$$

where r_{in} is the radius of thermal influence, t_c is the drilling fluid circulation time (at a given depth), r_w is the well radius, r_D is the dimensionless radial distance and γ is the thermal diffusivity of formation. Thus the dimensionless temperature in the wellbore and in formation at the end of mud circulation (at a given depth) can be expressed as:

$$T_{cD}(r_D,t_D) = \left\{ \begin{array}{ll} 1, & 0 \le r_D \le 1 \\ 1 - \dfrac{\ln r_D}{R_{in}}, & 1 \le r_D \le R_{in} \\ 0, & r_D > R_{in} \end{array} \right\}. \qquad (15\text{-}2)$$

To determine the temperature in the well ($r = 0$) after the circulation of fluid ceased, we used the solution of the diffusivity equation that describes cooling along the axis of a cylindrical body (Carslaw and Jaeger 1959) with known initial temperature distribution (Eq. (15-2)), placed in an infinite medium of constant temperature (Kutasov 1999). We obtained the following expression for T_{sD}

$$T_{sD} = \frac{T(0,t_s) - T_f}{T_w - T_f} = 1 - \frac{Ei(-pR_{in}^2) - Ei(-p)}{2 \ln R_{in}}, \quad t_D > 5, \tag{15-3}$$

$$p = \frac{1}{4nt_D}, \quad n = \frac{t_s}{t_c},$$

where t_s is the shut-in time.

It was assumed that for deep wells the radius of thermal influence is much larger than the well radius, and, therefore, the difference in thermal properties of drilling muds and formations can be neglected. In the analytical derivation of Eq. (15-3) two main simplifications of the drilling process were made: it was assumed that drilling is a continuous process and the effective mud temperature (at a given depth) is constant. For this reason field data were used to verify Eq. (15-3). Long-term temperature observations in deep wells of Russia, Belarus, and Canada were used for this purpose (Kutasov 1968; Djamalova 1969; Bogomolov et al. 1970; Kritikos and Kutasov 1988). The shut-in times for these wells covered a wide range (12 hours to 10 years) and the drilling time varied from 3 to 20 months. The observations showed that Eq. (15-3) gives a sufficiently accurate description of this process of temperature recovery in the borehole. In practice, for deep wells (large t_D and small p) we can assume that

$$R_{in} \approx D_0 \sqrt{t_D} \tag{15-4}$$

and

$$Ei(-p) \approx -\ln t_D - \ln n - \ln 4 + 0.5772. \tag{15-5}$$

Introducing Eqs. (15-4) and (15-5) into Eq. (15-3) yields:

$$\frac{T(0,t_s) - T_f}{T_w - T_f} = \frac{Ei\left(-\dfrac{D}{n}\right) + \ln n - D_1}{2 \ln t_D + 2 \ln D_0}, \tag{15-6}$$

where

$$D = D_0^2/4 = 1.1925, \quad D_1 = 0.5772 + \ln D = 0.7532.$$

If two measured shut-in temperatures (T_{s1}, T_{s2}) are available for the given depth with $t_s = t_{s1}$ and $t_s = t_{s2}$, we obtain

$$\frac{T_{s1} - T_f}{T_{s2} - T_f} = \frac{Ei\left(-\dfrac{D}{n_1}\right) + \ln n_1 - D_1}{Ei\left(-\dfrac{D}{n_2}\right) + \ln n_1 - D_1}. \tag{15-7}$$

Therefore:

$$T_f = T_{s2} - \gamma(T_{s1} - T_{s2}),$$ (15-8)

where

$$\gamma = \frac{Ei\left(-\dfrac{D}{n_2}\right) + \ln n_2 - D_1}{Ei\left(-\dfrac{D}{n_2}\right) - Ei\left(-\dfrac{D}{n_1}\right) + \ln \dfrac{n_2}{n_1}},$$ (15-9)

$$n_1 = \frac{t_{s1}}{t_c}, \quad n_2 = \frac{t_{s2}}{t_c}$$

Thus the well radius and thermal diffusivity of the formation have no influence on value of T_f, as the unknown parameters T_w and t_D have been eliminated. The quantities r_w and a, however, affect the value of T_f through T_{s1} and T_{s2}. The correlation coefficient γ monotonically changes with depth. As an example we present Table 15-1 for the well Rechitskaya 17-P (Rechitskaya oil field, Belarus), $t_{s1} = 1$ day and $t_{s2} = 209$ days.

T_f and T_w values were determined from observed temperatures at $t_s = 200$ days and 3210 days. At this depth determination of the formation (undisturbed) temperature with accuracy of 0.1°C requires a shut-in time of about eight years.

Table 15-1. Calculated formation temperatures, well Rechitskaya 17-P (Kutasov et al. 1971).

H, m	T_{s1}	T_{s2}	$-\gamma$	$\gamma(T_{s1} - T_{s2})$	T_f
500	24.9	24.21	0.215	−0.15	24.06
700	25.5	27.45	0.206	0.40	27.85
900	27.5	29.99	0.194	0.48	30.47
1100	29.35	32.84	0.183	0.64	33.48
1400	31.7	36.49	0.166	0.80	37.29
1600	33.8	39.09	0.154	0.81	39.90
1800	35.65	41.70	0.140	0.85	42.55
2000	37.4	44.32	0.128	0.88	45.20
2300	42.05	52.82	0.105	1.13	53.93
2600	46.0	57.49	0.082	0.95	58.44

Figure 15-1 presents the temperature versus shut-in time curve at the depth of Namskoe deep well (Yakutia republic, Russia; depth is 3003 m, total drilling time is 578 days).

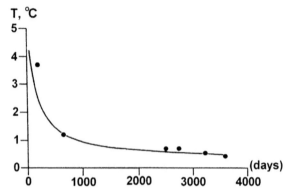

Figure 15-1. Rate of the temperature recovery at 500 depth in the Namskoe well (sand $a = 0.0020 \sim m^2h^{-1}$). Solid curve presents application of Eq. (15-3), and points are field measurements (Kutasov 1968).

Figure 15-2 presents the results of calculations of values T_f for the well 1225 (Kola Peninsula, Russia). Measured temperatures observed at $t_{s1} = 4.5$ days and

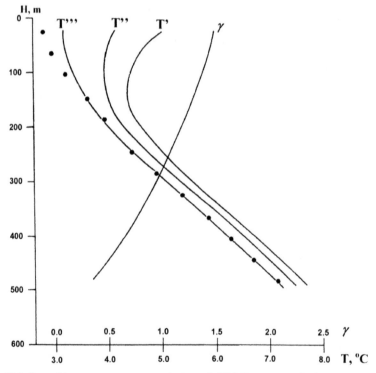

Figure 15-2. Rate of the temperature recovery in the well 1225. Termograms T'', T''', T' were observed at $t_s = 0.5$; 4.5; and 63 days correspondingly. Points designate the calculated values of T_f, and γ is the correlation coefficient (Kutasov 1968).

$t_{s2} = 20$ days were used (a total of seven temperature logs were made with $0.5 \leq t_s \leq 63$ days). The total drilling time of this well was 94 days, the depth 500 m.

The field data and the calculated T_f values show that, for a depth range 200–500 m, a shut-in time of two months is adequate if the accuracy in the determination of T_f is 0.03°C.

16

Determination of Formation Temperatures from Temperature Logs in Deep Boreholes: Comparison of Three Methods

The estimation of geothermal gradients, determination of heat flow density, well log interpretation, well drilling and completion operations, and evaluation of geothermal energy resources require knowledge of the undisturbed reservoir temperature. The drilling process, however, greatly alters the temperature of formation immediately surrounding the well. The temperature change is affected by several factors, such as the duration of drilling fluid circulation, the temperature difference between the reservoir and the drilling fluid, the well radius, the thermal diffusivity of the reservoir, and the drilling technology used. Given these factors, the exact determination of formation temperature at any depth requires a certain length of time in which the well is not in operation (Eppelbaum and Kutasov 2006b). In theory, this shut-in time is infinitely long to reach the original condition. There is, however, a practical limit to the time required for the difference in temperature between the well wall and surrounding reservoir to become a negligible quantity. The results of field and analytical investigations have shown that in many cases the effective temperature (T_w) of the circulating fluid (mud) at a given depth can be assumed constant during drilling or production (e.g., Lachenbruch and Brewer 1959; Ramey 1962; Edwardson et al. 1962; Jaeger 1961; Kutasov et al. 1966; Raymond 1969).

Here we should note that even for a continuous mud circulation process the wellbore temperature is dependent on the current well depth and other factors. The term "effective fluid temperature" is used to describe the temperature disturbance caused to formations while drilling. In their classical paper, Lachenbruch and Brewer (1959) have shown that the wellbore shut-in temperature mainly depends on the amount of thermal energy transferred to (or from) formations.

From the recent publications it is necessary to note a work of Zschocke (2005) where the correction function is defined by the drilling history, thermal parameters and at least two samples of the equilibrium temperature at different depths is needed.

Thus for every depth a value of T_w can be estimated from shut-in temperature logs. The objective of this chapter is to conduct a comparison of three methods: the Horner method (Dowdle and Cobb 1975; Jorden and Campbell 1984), the Three Point Method (Kutasov and Eppelbaum 2003), and the Two Logs Method (Kritikos and Kutasov 1988; Kutasov 1999) of predicting the undisturbed formation temperature from shut-in temperature logs in deep wells. Long-term temperature surveys in four–five wells were selected to estimate the formation temperatures and to compare measured and predicted transient formation temperatures.

16.1 Horner Method

The Horner method (HM) is widely used in petroleum reservoir engineering and in hydrogeological exploration to process the pressure-build-up test data for wells produced at a constant flow rate. Earlier we presented the main features and limitations of this method (please see Chapter 13).

16.2 Two Logs (Points) Method

The mathematical model of the Two Logs Method (TLM) is based on the assumption that in deep wells the effective temperature of drilling mud at a given depth can be assumed to be constant during the drilling process (please see Chapter 15).

16.3 Three Points Method

The main features of Three Points Method (TPM) and data processing technique were presented in Chapter 14.

16.4 Results of Calculations and Discussion

Temperature logs conducted in four wells are used to verify the suggested formula.

Extensive temperature measurements in the Northern Canada and Alaska were conducted by the Geothermal Service of Canada (Taylor and Judge 1976; Judge et al. 1981; Taylor et al. 1982) and by U.S. Geological Survey (Clow and Lachenbruch 1998). We selected long-term temperature surveys in four–five wells (Table 16-1) to estimate the formation temperatures and to compare measured and predicted transient formation temperatures by the given three methods. The results of several temperature surveys in the Well 167 are presented in Figure 16-1.

Table 16-1. Well data and references.

Well	Total vertical depth (m)	Drilling time (days)	Number of logs (n), Shut-in period (days)	Reference
No. 192, Kugpik D-13	3689	188	7 35–2835	Taylor et al. 1982
No. 275, Parsons N-17	3295	116	8 8–1192	Judge et al. 1981
No. 167, Unipkat I-22	4361	179	10 26–3042	Taylor et al. 1982
ADGO	2538	85	5 36–298	Taylor and Judge 1976
FCK, West Fish Creek No. 1, Alaska	735	73	6 227–2657	Clow and Lachenbruch 1998

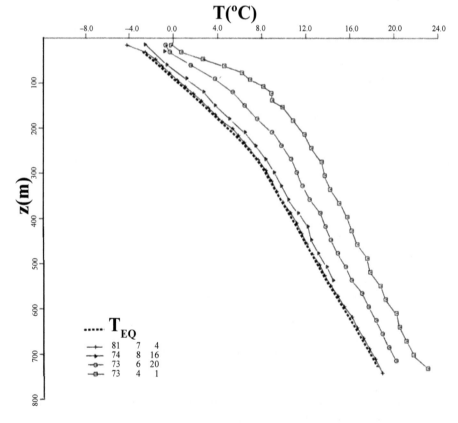

Figure 16-1. Observed temperature profiles in the well No. 167, measured on different dates (yy mm dd), T_{EQ}—the predicted temperature versus depth curve (Taylor et al. 1982).

For three wells (No. 167, 192 and 275) interpolation of observed temperature data was often used to refer temperatures to the same depth (Tables 16-2 and 16-3). In this section the subscripts '3', '*H*', and '2' refer to static (undisturbed) and transient temperatures obtained by TPM, HM and TLM techniques, respectively. R_3, R_H and R_2 are the squared average deviations.

Table 16-2. Observed shut-in temperatures (T_{01}, T_{02}, T_{03}) and formation temperatures calculated by three methods (see text for symbol explanation).

z, m	T_{01}, °C	T_{02}, °C	T_{03}, °C	T_{f3}, °C	T_{fH}, °C	T_{f2}, °C
\multicolumn{7}{c}{Well 192, t_{s1} = 35 d, t_{s2} = 128 d, t_{s3} = 320 d}
182.9	10.67	7.66	5.51	3.13	4.11	3.46
213.4	11.45	8.63	6.80	4.83	5.48	5.05
243.8	11.89	9.48	7.82	6.00	6.70	6.24
274.6	11.88	9.84	8.34	6.67	7.40	6.91
\multicolumn{7}{c}{Well 167, t_{s1} = 26 d, t_{s2} = 106 d, t_{s3} = 335 d}
304.8	13.72	11.29	9.62	8.34	8.72	8.47
335.3	14.21	11.86	10.50	9.50	9.62	9.56
365.8	15.03	12.61	11.30	10.35	10.40	10.40
396.2	15.76	13.46	12.07	11.03	11.22	11.11
426.7	16.15	13.95	12.64	11.67	11.83	11.74
457.2	16.66	14.50	13.25	12.34	12.45	12.40
487.7	17.55	15.18	14.04	13.24	13.16	13.27
518.2	17.83	15.83	14.58	13.65	13.86	13.73
\multicolumn{7}{c}{Well 275, t_{s1} = 8 d, t_{s2} = 88 d, t_{s3} = 190 d}
426.7	10.04	5.36	4.08	2.43	3.13	2.58
457.2	11.03	6.35	4.92	3.04	4.01	3.25
487.7	11.79	7.04	5.73	4.05	4.78	4.21
518.5	12.50	7.78	6.51	4.89	5.56	5.04

We should also note that for the Two Logs Method logs $t_s = t_{s2}$ and $t_s = t_{s3}$ were utilized. A minimum of two temperature logs is required to utilize the HM and TLM techniques to determine the undisturbed formation temperature (T_f). However, it is preferable that an additional temperature log be available to estimate the accuracy of the T_f predictions. Respectively, to use the TPM technique one more temperature log is required. In this case, the temperature differences (at a given depth) should be significantly larger than the absolute accuracy of temperature measurements. A good example is the result of temperature surveys conducted in the AGDO well, Alaska (Table 16-4).

Table 16-3. Comparison between measured shut-in temperatures (T_{ob}) and calculated by three methods transient temperatures.

z, m	Well 192, $t_s = 660$ d				Well 192, $t_s = 1755$ d			
	T_{ob}, °C	T_3, °C	T_H, °C	T_2, °C	T_{ob}, °C	T_3, °C	T_H, °C	T_2, °C
182.9	4.16	4.42	4.99	4.54	3.64	3.65	4.46	3.88
213.4	5.48	5.89	6.28	5.96	4.99	5.26	5.81	5.41
243.8	6.81	6.98	7.40	7.06	6.39	6.39	6.98	6.56
274.6	7.54	7.57	8.00	7.66	7.17	7.03	7.64	7.21
R, °C	$R_3 = 0.26\ R_H = 0.69\ R_2 = 0.34$				$R_3 = 0.15\ R_H = 0.69\ R_2 = 0.26$			
	Well 167, $t_s = 528$ d				Well 167, $t_s = 868$ d			
304.8	9.28	9.20	9.41	9.24	8.97	8.89	9.16	8.95
335.3	9.94	10.17	10.25	10.19	9.58	9.92	10.02	9.96
365.8	10.66	10.99	11.02	11.00	10.28	10.75	10.80	10.77
396.2	11.57	11.72	11.84	11.75	11.19	11.47	11.61	11.51
426.7	12.28	12.32	12.41	12.34	11.80	12.08	12.20	12.12
457.2	12.72	12.95	13.02	12.96	12.41	12.73	12.82	12.75
487.7	13.43	13.77	13.74	13.78	13.13	13.57	13.53	13.59
518.2	14.14	14.27	14.40	14.30	13.83	14.04	14.20	14.09
R, °C	$R_3 = 0.22\ R_H = 0.27\ R_2 = 0.23$				$R_3 = 0.32\ R_H = 0.40\ R_2 = 0.35$			
	Well 275, $t_s = 333$ d				Well 275, $t_s = 703$ d			
426.7	3.41	3.45	3.82	3.49	2.75	2.94	3.48	3.03
457.2	4.23	4.21	4.72	4.26	3.61	3.63	4.37	3.75
487.7	5.08	5.09	5.48	5.13	4.48	4.57	5.13	4.66
518.5	5.94	5.89	6.26	5.93	5.31	5.39	5.91	5.48
R, °C	$R_3 = 0.03\ R_H = 0.41\ R_2 = 0.05$				$R_3 = 0.11\ R_H = 0.69\ R_2 = 0.20$			

Table 16-4. Observed shut-in temperatures and calculated formation temperatures, well ADGO.

z, m	$t_{s1} = 36$ d	$t_{s2} = 124$ d	T_{f1}, °C	T_{f2}, °C
201.5	8.78	6.83	5.40	5.49
241.1	8.72	7.11	5.93	6.01
280.7	9.50	7.78	6.53	6.62
320.3	9.94	8.33	7.17	7.25
360.0	11.00	9.22	7.94	8.03
399.6	10.78	9.44	8.48	8.55
439.2	11.78	10.39	9.41	9.48
478.8	12.78	11.33	10.31	10.38
518.5	13.72	12.22	11.17	11.25

In this well five temperature logs were conducted at $t_s = 36, 124, 132, 267$ and 298 days. For the 201.5–518.5 m section of the well the temperature differences for two logs conducted at $t_s = 124$ and 132 days is 0.05–$0.11°C$. Now let us assume that only three temperature logs are available. It is obvious that in this case the Three Points Method cannot be used to determine the values of T_f. The HM and TLM techniques were used to calculate values of T_f (Table 16-4). After this the values of T_f were used to calculate the transient temperatures for logs at $t_s = 267$ and 298 days (Table 16-5). For the estimation of the accuracy of the T_f predictions we compared the calculated transient temperatures with observed ones (Tables 16-3 and 16-5) and computed the values of average squared temperature deviations, ΔT. From Tables 16-3 and 16-5 it follows that the values of ΔT are practically the same and both methods can predict the values of T_f with the accuracy of about $0.1°C$.

Table 16-5. Comparison between measured shut-in temperatures (T_{ob}) and calculated by two methods transient temperatures, well ADGO.

z, m	Shut-in time, 267 days			Shut-in time, 298 days		
	T_o, °C	T_H, °C	T_2, °C	T_o, °C	T_{Ho}, °C	T_2, °C
201.5	6.00	6.15	6.17	5.94	6.08	6.11
241.1	6.39	6.55	6.57	6.39	6.49	6.52
280.7	7.06	7.18	7.21	7.00	7.12	7.14
320.3	7.61	7.78	7.80	7.56	7.72	7.74
360.0	8.50	8.61	8.63	8.44	8.55	8.57
399.6	8.83	8.98	9.00	8.78	8.94	8.96
439.2	9.72	9.41	9.94	9.72	9.87	9.89
478.8	10.72	10.84	10.86	10.67	10.79	10.81
518.5	11.56	11.72	11.74	11.50	11.66	11.69
R, °C	$R_H = 0.17\ R_2 = 0.17$			$R_H = 0.14\ R_2 = 0.16$		

Table 16-6. Observed shut-in temperatures and calculated by three methods formation temperatures, well FCK.

z, m	T_{o1}, °C $t_{s1} = 227$ d	T_{o2}, °C $t_{s2} = 506$ d	T_{o3}, °C $t_{s3} = 867$ d	T_{β}, °C	$T_{\beta P}$, °C	T_{f2}, °C
368.8	5.86	4.97	4.56	4.02	4.11	3.98
399.3	6.82	6.17	5.74	5.12	5.41	5.14
432.8	7.98	7.35	6.95	6.39	6.64	6.34
478.5	9.66	9.02	8.59	7.94	8.28	7.97
512.1	10.65	10.03	9.65	9.10	9.35	9.11
542.6	12.01	11.44	11.07	10.52	10.79	10.54
573.0	13.02	12.43	12.11	11.67	11.83	11.66
606.6	14.19	13.62	13.31	12.88	13.03	12.87
640.1	15.23	14.84	14.55	14.12	14.37	14.15
670.6	16.36	15.84	15.57	15.21	15.33	15.19
704.1	17.94	17.43	17.14	16.73	16.91	16.73

For four wells all three methods were used to compute the formation temperatures and to calculate the transient temperatures at two shut-in times (Tables 16-3 and 16-6). For the wells No. 167 (Figure 16-2) and No. 275 (for four depths, Table 16-3) the Three Point Method provides the best estimate of the formation and transient temperatures. Interesting results are obtained for well No. 167 (Table 16-3). In this case the transient temperatures computed by three methods (at $t_s = 528$ days and $t_s = 868$ days) are very close. However, the deviations of calculated values from measured ones (expressed in terms of ΔT) are relatively large. To some extent this can be attributed to the accuracy of field data and to the interpolation method used (to refer observed temperatures to the same depth). However, even in this case the Three Point Method gives a better estimate of formation and transient temperatures than the Horner and the Two Logs Methods. Comparison of predicted undisturbed formation temperatures (Table 16-32) and those T_{EO} (Figure 16-1), determined from 9 long-term temperature logs (Taylor et al. 1982, p. 35 and Figure 3) are in a fair agreement. The authors used the empirical Lachenbruch-Brewer formula (Horner method) to estimate the undisturbed formation temperature (T_{EO}). The formation temperatures computed by TPM and TLM techniques for the FCK Well are in a very good agreement, while the Horner method provides slightly higher values (Table 16-7) (Eppelbaum and Kutasov 2006b).

Analysis of Table 16-7 and Figure 16-3 testifies that both methods (TPM and TLM) can be used with a high accuracy to predict the transient formation temperatures. An example of field data processing by the "PERMTEMP" program is presented in Table 16-8.

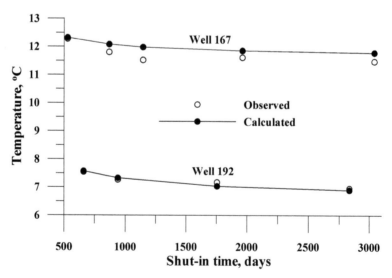

Figure 16-2. Comparison between measured shut-in and calculated by Three Point Method temperatures at two depths: $z = 274.6$ m (well 192); $z = 426.7$ m (well 167).

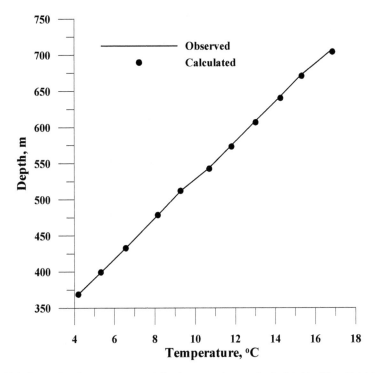

Figure 16-3. Comparison between measured shut-in temperatures and calculated by Three Point Method, well FCK, $t_s = 2657$ days.

Table 16-7. Comparison between measured shut-in temperatures and calculated by three methods transient temperatures, well FCK (West Fish Creek No.1, Alaska).

z, m	$t_s = 1576$ d			
	T_{ob}, °C	T_3, °C	T_H, °C	T_2, °C
368.8	4.35	4.31	4.38	4.30
399.3	5.50	5.46	5.63	5.47
432.8	6.73	6.70	6.85	6.70
478.5	8.33	8.30	8.49	8.31
512.1	9.46	9.40	9.55	9.41
542.6	10.86	10.82	10.98	10.83
573.0	11.93	11.91	12.01	11.91
606.6	13.13	13.11	13.21	13.11
640.1	14.33	14.36	14.50	14.37
670.6	15.36	15.40	15.48	15.40
704.1	16.93	16.95	17.06	16.95
R, °C	$R_3 = 0.035 \; R_H = 0.115 \; R_2 = 0.033$			

Table 16-8. Predicted formation temperatures, well FCK; $t_{s1} = 227$, $t_{s2} = 506$, and $t_{s3} = 867$ days.

z, m	t_{xD}	A, 1/day	R_x	T_f°C
30.5	1.01	0.003405	3.196	−8.90
61.0	1.83	0.006212	3.951	−8.37
91.4	1.16	0.003970	3.347	−7.76
112.8	0.65	0.002250	2.761	−7.06
146.3	0.50	0.001740	2.539	−5.62
179.8	0.62	0.002191	2.717	−3.91
368.8	1.14	0.004342	3.335	4.02
399.3	0.69	0.002632	2.808	5.12
432.8	0.71	0.002771	2.843	6.39
478.5	0.61	0.002431	2.711	7.94
512.1	0.68	0.002720	2.798	9.10
542.6	0.64	0.002579	2.740	10.52
573.0	0.79	0.003248	2.941	11.67
606.6	0.78	0.003271	2.934	12.88
640.1	0.51	0.002144	2.555	14.12
670.6	0.78	0.003359	2.934	15.21
704.1	0.67	0.002898	2.783	16.73

In our study we did not take into account the effect of the geothermal gradient on the restoration of the natural temperature field of formations. This may have some effect on the predicted values of T_f in regions with high geothermal gradients.

We can conclude that the TPM can be successfully used for determination of static and transient formations temperatures. The presented field examples show that TPM has an advantage when compared with the Horner method and the Two Logs Method. For deep wells we recommend conducting at least four temperature surveys. The advantage of the Three Point Method is that TPM can be applied for processing of temperature logs conducted in permafrost areas. In this case (in the permafrost section of the well) the starting point in the well thermal recovery is moved from the end of well completion to the moment of time when the refreezing of formations were completed (Kutasov and Eppelbaum 2003, 2010).

17

Geothermal Temperature Gradient

Most of temperature surveys are conducted in boreholes. The total vertical depth of the boreholes (< 10 km) is small in comparison with the radius of the Earth (6371 km) and for this reason the curvature of the Earth's surface can be neglected. It is a known fact that the formation temperature increases with depth. Only in some offshore-onshore transition areas and permafrost regions at shallow depths (several hundreds of meters) the temperature reduces with depth. The rate of the temperature increase is determined by the geothermal gradient (Γ). In a general case the geothermal gradient has three components (Eppelbaum et al. 2014),

$$\Gamma = \sqrt{\Gamma_x^2 + \Gamma_y^2 + \Gamma_z^2}, \quad \Gamma_x = \frac{\partial T}{\partial x}, \quad \Gamma_y = \frac{\partial T}{\partial y}, \quad \Gamma_z = \frac{\partial T}{\partial z}. \tag{17-1}$$

The value of HFD determines the amount of heat per unit of area and per unit of time which is transmitted by heat conduction from the Earth's interior. For isotropic and homogeneous formations, where coefficient of thermal conductivity (λ) is a constant, the value of heat-flow density, HFD (q means vector), can be calculated from the Fourier equation:

$$\mathbf{q} = -\lambda \, grad \, T. \tag{17-2}$$

The minus sign shows that heat flows from points with high temperatures to points with lower temperatures. Thus to calculate the HFD we need to estimate the value of the static (undisturbed) geothermal gradient (Γ) and to measure the thermal conductivity of the formation. Two methods of combining thermal conductivity and temperature gradient data are used: interval method and the Bullard method (Powell et al. 1988). It is assumed that the effect of climatic changes, relief, underground water movement, subsurface conductivity variations on the temperature gradient (G) have been estimated and the corrected value of G is close to the value of Γ. In the interval method, for each depth interval a temperature gradient is combined with the representative value of formation thermal conductivity.

The temperature gradient is a differential quantity hence the process by which the geothermal (undisturbed) gradient (Γ) is restored is distinct from temperature recovery process. Now the question arises: how accurately is the value of Γ determined from temperature measurements taken at a short time after cessation of drilling.

The determination of the geothermal gradient in the 407–467 m depth interval in well 1225 (Figure 17-1) showed that the geothermal gradient can be determined with an accuracy of 10% just after two days after drilling ceases, and with an accuracy of 3% after 20 days (Kutasov 1999).

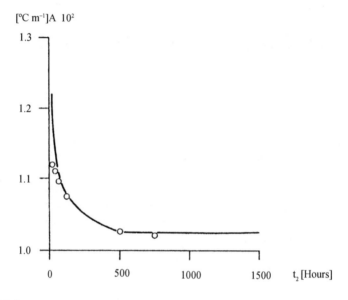

Figure 17-1. Temperature gradient recovery in the 407–467 m section of the well 1225 (depth 500 m the duration of drilling 94 days, a = 0.0027 m²h⁻¹).

Solid curve present the calculated values, and open circles—field data (Kutasov 1999).

In the depth range of 500–700 m in the Namskoe deep well (Yakutia, Russia; depth is 3003 m, total drilling time is 578 days) the geothermal gradient can be estimated with an accuracy of 6% after only after 50 days of shut-in (Kutasov 1999).

On the basis of Eq. (15-8) we will derive a relationship linking the geothermal gradient with the transient vertical temperature gradients.

$$T_f = T_{s2} - \gamma(T_{s1} - T_{s2}),$$

where T_f is the formation temperature, T_{s1}, T_{s2} are temperatures (at a given depth) obtained at shut-in times $t_s = t_{s1}$ and $t_s = t_{s2}$. The function γ was defined earlier (Eq. (15-9)). On the basis of Eq. (15-8) we will derive a relationship linking the geothermal gradient with the transient vertical temperature gradients.

Let us assume that two temperature logs.

$T_{s1} = T(0,t_{s1})$ and $T_{s2} = T(0,t_{s2})$ are available and the parameters for two points (depths) are:

$$H_2 = H_1 + \Delta H, \quad t_{c2}, \quad H_1, \quad t_{c1},$$

where t_{c1} and t_{c2} are time of drilling fluid circulation at two depths.

Then for the depths H_2 and H_1

$$T_{f2} = T_{2s2} - \gamma_2 \left(T_{2s1} - T_{2s2} \right), \tag{17-3}$$

$$T_{f1} = T_{1s2} - \gamma_1 \left(T_{1s1} - T_{1s2} \right), \tag{17-4}$$

where T_{f2} and T_{f1} are the formation temperatures at depths H_2 and H_1, T_{2s1}, T_{2s2}, T_{1s1}, T_{1s2} are shut-in temperatures at $t_s = t_{s1}$ and $t_s = t_{s2}$.

The corresponding values of γ function are

$$\gamma_2 = \frac{Ei\left(-\dfrac{D}{n_2}\right) + \ln n_{22} - D_1}{Ei\left(-\dfrac{D}{n_{22}}\right) - Ei\left(-\dfrac{D}{n_{12}}\right) + \ln \dfrac{n_{22}}{n_{12}}}, \tag{17-5}$$

$$n_{12} = \frac{t_{s1}}{t_{c2}}, \quad n_{22} = \frac{t_{s2}}{t_{c2}},$$

$$\gamma_1 = \frac{Ei\left(-\dfrac{D}{n_{12}}\right) + \ln n_{12} - D_1}{Ei\left(-\dfrac{D}{n_{12}}\right) - Ei\left(-\dfrac{D}{n_{11}}\right) + \ln \dfrac{n_{12}}{n_{11}}}, \tag{17-6}$$

$$n_{11} = \frac{t_{s1}}{t_{c1}}, \quad n_{12} = \frac{t_{s2}}{t_{c1}}.$$

The values of the geothermal temperature gradient (Γ) and transient temperature gradients (A_1 and A_2) are

$$\Gamma = \frac{T_{f2} - T_{f1}}{\Delta H}, \quad A_1 = \frac{T_{2s1} - T_{1s1}}{\Delta H}, \quad A_2 = \frac{T_{2s2} - T_{1s2}}{\Delta H}.$$

Example of calculations. Well 192 (Taylor et al. 1982, see also Table 16-1).

Total vertical depth is 3689.0 m, and total drilling time is 188.0 days. We will estimate the geothermal gradient in the well section 274.6–396.5 m. We selected $t_{s1} = $ 320 days and $t_{s2} = 1775$ days and observed shut-in temperatures were: $T_{2s1} = 11.25°C$, $T_{2s2} = 10.15°C$, $T_{1s1} = 8.83°C$, $T_{1s2} = 7.17°C$.

The calculated parameters are:

$$\gamma_1 = -0.2562, \qquad \gamma_2 = -0.2548, \qquad T_{f1} = 6.8704\,^\circ\mathrm{C} \qquad T_{f2} = 9.8610^\circ\mathrm{C}$$

and values of Γ, A_1 and A_2 are

$$\Gamma = \frac{9.8610 - 6.8704}{396.5 - 274.6} = 0.02453\,(^\circ\mathrm{C/m}), \quad A_1 = \frac{11.25 - 8.34}{396.5 - 274.6} = 0.02387(^\circ\mathrm{C/m}),$$

$$A_2 = \frac{10.15 - 7.17}{396.5 - 274.6} = 0.02445(^\circ\mathrm{C/m}).$$

We can also estimate the ratios

$$\frac{A_1}{\Gamma} = \frac{0.02387}{0.02453} = 0.973, \qquad \frac{A_2}{\Gamma} = \frac{0.02445}{0.02453} = 0.997.$$

Thus after 320 days of shut-in the geothermal temperature gradient can be estimated with 3% of accuracy.

18

Estimation of the Geothermal Gradients from Single Temperature Log-Field Cases

It was shown earlier (Kutasov 1999) that at least two transient temperature surveys are needed to determine the geothermal gradient with an adequate accuracy. However, in many cases only one temperature log is conducted in a shut-in borehole (Kutasov and Eppelbaum 2009). For these cases we propose an approximate method for estimation of the geothermal gradient. The utilization of this method is demonstrated on four field examples.

Knowledge of the geothermal temperature profiles is also needed to increase the accuracy of electric and temperature log interpretation. The drilling process greatly disturbs the temperature of the formations around the wellbore. This temperature change is affected by several parameters and technical choices, such as the duration of drilling fluid circulation, the temperature difference between the formations and drilling fluid, the well radius, the thermal diffusivity of formations, and the drilling technology used. Given these factors, the exact determination of formations temperatures and geothermal gradients requires a certain length of time in which the well is not in operation. In theory, this shut-in time is infinitely long. There is, however, a practical limit to the time required for the difference in temperature between the well wall and the surrounding rocks to become vanishingly small. The key point is the observation that the temperature gradient is a differential quantity hence the process by which the geothermal (undisturbed) gradient is restored is distinct from the temperature recovery process (Kutasov and Eppelbaum 2015). For example, in the depth range of 500–700 m in the Namskoe deep well (Yakutia, Russia; depth 3003 m, total drilling time 578 days) geothermal gradient was estimated with an accuracy of 6% after only 50 days of shut-in (Kutasov 1999).

18.1 Temperature Disturbance of Formations

In many cases (due to technical, financial and other limitations) only one temperature log is conducted in a shut-in borehole. For these cases we suggested an approximate method for estimation of the geothermal gradient (Kutasov and Eppelbaum 2009). It was shown that the thermal effect of drilling operations (at a given depth) could be approximated by a constant (effective) drilling mud temperature (Lachenbruch and Brewer 1959; Jaeger 1961; Kutasov 1968, 1999). For a constant well wall temperature a solution which describes the dimensionless temperature distribution T_D in formation surrounding the wellbore during the fluid circulation is known but is expressed through a complex integral (Jaeger 1956). By introducing the adjusted dimensionless circulation time t_D^* we have found (Kutasov 1987, 1999; see also Chapter 2) that the exponential integral (a tabulated function) can be used to approximate function T_D. For $r_D \geq 1$

$$T_D = \frac{T(r, t_c) - T_f}{T_c - T_f} = \frac{Ei(-\beta r_D^2)}{Ei(-\beta)}, \tag{18-1a}$$

$$t_D = \frac{at_c}{r_w^2}, r_D = \frac{r}{r_w}, \beta = \frac{1}{4t_D^*}, \quad Ei(x) = \int_x^\infty \frac{e^{-t}}{t} dt,$$

$$t_D^* = t_D \left[1 + 1/(1 + AF)\right], t_D \leq 10,$$

$$F = \left[\ln(1 + t_D)\right]^n; \quad n = 2/3, \tag{18-1b}$$

$$A = 7/8.$$

For $t_D > 10$

$$t_D^* = t_D \frac{\ln t_D - \exp\left(-0.236\sqrt{t_D}\right)}{\ln t_D - 1}, \tag{18-1c}$$

where r is the radial distance (at well axis $r = 0$), t_c is the drilling mud circulation time at a given depth, r_D is the dimensionless distance, r_w is the radius of the well, t_D is the dimensionless circulation time, $Ei(x)$ is the exponential integral, T_c is the circulation mud temperature, T_f is the formation (undisturbed) temperature, $T(r, t_c)$ is the radial temperature, $T_D(r_D, t_D)$ is the dimensionless radial temperature, and a is the thermal diffusivity of formations.

Zschocke (2005) used Eqs. (18-1b) and (18-1c) to develop a method allowing to correct the entire temperature log when at least two data sets are available.

18.2 Wellbore Shut-in Temperature

Using the initial temperature distribution (Eq. (18-1a)) we obtained the following expression for the wellbore (at $r = r_w$) shut-in temperature T_s (Kutasov 1987, 1999):

$$\gamma = T_{sD} = 1 - \frac{Ei\left[-\beta\left(1 + \frac{t_{D}^{*}}{t_{sD}}\right)\right]}{Ei(-\beta)},$$ (18-2)

$$T_{sD} = \frac{T_{s} - T_{f}}{T_{c} - T_{f}}, \quad t_{sD} = \frac{at_{s}}{r_{w}^{2}},$$

where t_{s} is the shut-in time and t_{sD} is the dimensionless shut-in time.

The derivation of Eq. (18-2) assumes that the difference in thermal properties between the drilling mud and formation is negligible. Although this is a conventional assumption even for interpreting bottom-hole temperature surveys, when the circulation periods are small, Eq. (18-2) should be used with caution for very small shut-in times. Now by using Eq. (18-2) we can show why the recovery of the geothermal gradient is distinct from temperature recovery process. Let's assume that for depth h_{1} and h_{2} the formation temperatures are T_{1} and T_{2} and shut temperatures are T_{s1} and T_{s2}. Let as also assume that for a short section of the well the temperature of the drilling fluid (T_{c}) and the value of T_{sD} are constants. Then from Eq. (18-2) follows

$$\begin{cases} T_{sD}\left(T_{c} - T_{1}\right) = T_{s1} - T_{1} \\ T_{sD}\left(T_{c} - T_{2}\right) = T_{s2} - T_{2} \end{cases}$$

or

$$T_{sD}\left(T_{2} - T_{1}\right) = -\left(T_{s2} - T_{s1}\right) + \left(T_{2} - T_{1}\right)$$

or

$$\begin{cases} T_{sD}\Gamma = -\Gamma_{1} + \Gamma, \\ \Gamma = \dfrac{T_{2} - T_{1}}{h_{2} - h_{1}}, \quad \Gamma_{1} = \dfrac{T_{s2} - T_{s1}}{h_{2} - h_{1}} \end{cases}.$$

It should be noted that with reduction of the value of T_{sD}, $\Gamma_{1} \rightarrow \Gamma$.

18.3 Prediction of the Geothermal Gradient

For deep wells $T_{sD} = f(t_{D}, t_{sD})$ and for small section of the well we can assume that the values of T_{c} and T_{sD} are *constants*. Then the formation temperature at some depth h is

$$T = T_{1} + \Gamma\left(h - h_{1}\right), \quad h_{2} \geq h > h_{1}.$$ (18-3)

Combining Eqs. (18-2) and (18-3) we obtain the following linear equation

$$T_{s} = \Gamma h(1 - \gamma) + B, \quad B = T_{1} + \gamma\left(T_{c} - T_{1}\right).$$ (18-4)

To estimate the value γ we suggest the following procedure: calculate the average vertical depth $h_{aver} = \dfrac{h_1 + h_2}{2}$ and then estimate the disturbance time (mud circulation time) for this depth

$$t_c \approx t_t - t_{aver},$$

where t_t is the total drilling time, t_{aver} is period of time needed to reach the depth h_{aver}. The value of t_{aver} can be determined from the drilling records. When drilling records are not available the following formula can be used:

$$t_c = t_t\left(1 - \frac{h_{aver}}{H}\right),$$

where H is the total vertical depth of the well. To determine dependable values of Γ and B from Eq. (18-4) we utilized a linear regression program. For 80–120 m well sections we used 4–5 pairs of temperature depth data (see Tables 18-1–18-4). To speed up calculations we prepared a computer program which was utilized to process field data.

Table 18-1. The results of calculations of Γ and Γ^*. Well 272 (Parsons L-43).

	t_s = 82 days					t_s = 29.4 months		
h, m	T_{sD}	T, °C	a, m²/hr	Γ, °C/m	Γ^*, °C/m	h, m	T, °C	Γ^*, °C/m
365.8	0.0934	5.73	0.0030	0.03106	0.02789	367.6	2.62	0.03276
396.2	0.0928	6.71	0.0040	0.03086		428.5	4.44	
426.7	0.0922	7.71	0.0050	0.03072		487.7	5.09	
457.2	0.0918	8.29				459.3	5.85	
487.7	0.0911	9.19				490.1	6.49	

18.4 Field Examples

Example 1. Well 272 (Parsons L-43) was drilled for 53 days to the total vertical depth of 3305 m and the well diameter is 0.438 m (Judge et al. 1981). The results of calculations after formulas 18-2–18-4 are presented in Table 18-1.

Here and below Γ is the geothermal gradient obtained from Eq. (18-4) and Γ^* is the temperature gradient estimated directly from measured data. In both cases the linear regression was used to process field data. From Table 18-1 follows that the values of Γ and Γ^* (the last column) are very close.

For this reason the temperature observations conducted after 82 days of shut-in in order to obtain more accurate value of the geothermal gradient are not necessary. At the same time the temperature profiles at t_s = 82 days and t_s = 29.4 months are quite different.

Example 2. Well 193 (Ikhil I-37) was drilled for 237 days to the total vertical depth 4704 m, and the well diameter is 0.438 m (Taylor et al. 1982). The results of calculations after Eqs. (18-2)–(18-4) are presented in Table 18-2. From Table 18-2 follows the values of Γ and Γ^* (the last column) are very close. Therefore the shut-in time of 62 days is sufficient to obtain the value Γ with a good accuracy.

Table 18-2. The results of calculations of Γ and Γ^*. Well 193 (Ikhil I-37).

	Shut-in time 62 days					Shut-in time 55.5 months		
h, m	T_{sD}	T, °C	a, m²/h	Γ, °C/m	Γ^*, °C/m	h, m	T, °C	Γ^*, °C/m
425.5	0.2256	7.89	0.0030	0.03178	0.02417	426.7	3.26	0.03114
456.0	0.2251	8.09	0.0040	0.03134		457.2	4.17	
486.5	0.2245	9.15	0.0050	0.03103		487.7	5.09	
516.9	0.2239	9.98				517.9	6.11	

Example 3. Well 175 (Gemini E-10) was drilled for 145 days to the total vertical depth of 3845 m and the well diameter is 0.438 m (Judge et al. 1981). The results of calculations after Eqs. (18-2)-(18-4) are presented in Table 18-3.

From Table 18-3 follows the values of Γ and Γ^* (at $t_s = 440$ days and at $t_s = 795$ days) are very close. Therefore the shut-in time of 53 days is sufficient to obtain the value Γ with a good accuracy.

Table 18-3. The results of calculations of Γ and Γ^*. Well 175 (Gemini E-10).

	$t_s = 53$ days				$t_s = 440$ days		$t_s = 795$ days	
	$\Gamma^* = 0.03293$/m		a, m²/h	Γ, °C/m	$\Gamma^* = 0.04003$°C/m		$\Gamma^* = 0.04086$°C/m	
h, m	T_{sD}	T, °C			h, m	T, °C	h, m	T, °C
680.6	0.1969	15.79	0.0030	0.04182	691.3	11.38	675.1	10.43
711.4	0.1961	16.66	0.0040	0.04129	713.9	12.26	705.9	11.70
741.9	0.1953	17.61	0.0050	0.04092	736.4	13.05	736.1	12.84
772.4	0.1946	18.79			759.0	14.00	766.6	14.13
802.7	0.1938	19.75			781.5	14.86	797.1	15.44
					804.1	15.87		

Example 4. Well Imperial ADGO P-25 was drilled for 85 days to the total vertical depth of 2538 m and the well diameter is 0.438 m (Taylor and Judge 1976). The results of calculations after formulas 18-2–18-4 are presented in Table 18-4. From Table 18-4

Table 18-4. The results of calculations of Γ and Γ^*. Well Imperial ADGO P-25.

h, m	$t_s = 36$ days $\Gamma^* = 0.00945$°C/m $\Gamma = 0.01225$°C/m		$t_s = 125$ days $\Gamma^* = 0.01187$°C/m $\Gamma = 0.01303$°C/m		$t_s = 298$ days $\Gamma^* = 0.01268$°C/m $\Gamma = 0.01326$°C/m	
	T, °C	T_{sD}	T, °C	T_{sD}	T, °C	T_{sD}
201.5	8.78	0.2070	6.83	0.0908	5.94	0.0445
221.3	8.56	0.2063	6.94	0.0904	6.22	0.0443
241.1	8.72	0.2056	7.11	0.0899	6.39	0.0440
260.9	9.00	0.2049	7.39	0.0895	6.61	0.0437
280.7	9.50	0.2042	7.78	0.0890	7.00	0.0435

follows that the accuracy of predicted values of the geothermal gradient increases with the reduction of values of T_{sD}.

As we can see, the values of T_{sD} in all four sections (Tables 18-1–18-4) are indeed practically constant. The second requirement for utilization of the suggested method is that the formation coefficient of thermal conductivity should be assumed as a constant within the well's section $h_2 - h_1$. For this reason the lithological profile of the well should be available. To select an interval $h_2 - h_1$ it is useful to have a graph T_{sD} versus depth (Eq. (18-2)). For three wells the function $T_{sD} = f(z)$ is presented in Figure 18-1. Table 18-5 also shows that the function $T_{sD} = f(z)$ slowly changes with depth.

Table 18-5. The values of T_{sD} for two sections of three wells: well 193 ($t_s = 62$ days), well 272 ($t_s = 82$ days), well 175 ($t_s = 53$ days); $a = 0.0050$ m²/hr.

z, m	Well number		
	175	193	272
200	0.2079	0.2296	0.0964
250	0.2068	0.2288	0.0955
300	0.2057	0.2279	0.0946
350	0.2046	0.2270	0.0937
400	0.2035	0.2261	0.0927
450	0.2023	0.2252	0.0918
500	0.2012	0.2243	0.0908
550	0.2000	0.2233	0.0899
600	0.1988	0.2224	0.0889
2600	0.1286	0.1713	0.0374
2650	0.1259	0.1695	0.0356
2700	0.1231	0.1676	0.0337
2750	0.1202	0.1658	0.0318
2800	0.1173	0.1639	0.0299
2850	0.1142	0.1619	0.0279
2900	0.1110	0.1599	0.0258
2950	0.1077	0.1579	0.0237
3000	0.1043	0.1558	0.0215

Thus an approximate method of determination of the geothermal gradient is proposed. The main preference of this method is that only one temperature log is needed to apply this procedure. This method can be used to estimate the duration of shut-in time required to obtain the value of geothermal gradient with (by evaluation of the T_{sD} function at various values of shut-in time) a suitable accuracy.

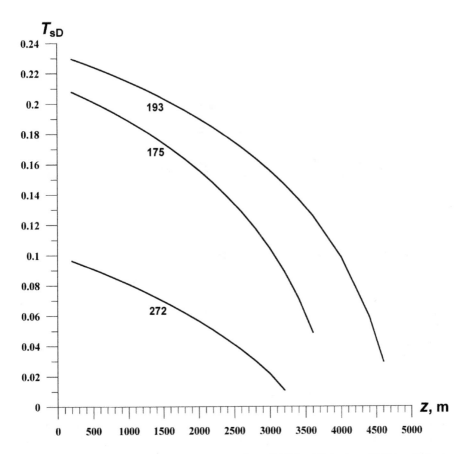

Figure 18-1. The function T_{sD} versus depth for three wells: well 193 (t_s = 62 days), well 272 (t_s = 82 days), well 175 (t_s = 53 days); a = 0.0050 m²/hr.

19

Radial Temperature Distribution

19.1 Radial Temperature Distribution during Drilling

To determine the temperature distribution $T(r, t)$ in formations near a wellbore with a constant bore-face temperature it is necessary to obtain a solution of the diffusivity equation for the following boundary and initial conditions:

$$\left\{ \begin{array}{ll} T(r,0) = T_f, & r_w \leq r \leq \infty, \quad t > 0, \\ T(r_w,t) = T_w, & T(\infty,t) = T_f \end{array} \right\}.$$

It is well known that in this case the diffusivity equation has a solution in a complex integral form (Jaeger 1956; Carslaw and Jaeger 1959). Jaeger (1956) presented results of a numerical solution for the dimensionless temperature $T_D(r_D, t_D)$ with values of r_D ranging from 1.1 to 100 and t_D ranging from 0.001 to 1000. We have found that the exponential integral (a tabulated function) can be used to describe the temperature field of formations around a well with a constant bore-face temperature (Kutasov 1999)

$$T_D(r_D, t_D) = \frac{T(r,t) - T_f}{T_w - T_f} = \frac{Ei\left(-\dfrac{r_D^2}{4t_D^*}\right)}{Ei\left(-\dfrac{1}{4t_D^*}\right)}, \tag{19-1}$$

$$t_D = \frac{at}{r_w^2}, \quad r_D = \frac{r}{r_w}, \quad t_D^* = Gt_D,$$

where a is the thermal diffusivity of formations, t is the time, r_w is the well radius, r is the radial coordinate, G is the correlation coefficient (see Chapter 2), T_w is the temperature of the drilling fluid (at a given depth), t_D^* is the adjusted dimensionless circulation time and Ei is the exponential integral. In Table 19-1 values of T_D calculated after Eq. (19-1) and results of a numerical solution are compared. The agreement between values of T_D calculated by these two methods is seen to be good.

Table 19-1. Dimensionless temperature $T_D(r_D, t_D)$ for a well with constant bore-face temperature, first line is obtained from Eq. (19-1); second line-numerical solution (Jaeger 1956).

t_D	$T_D \cdot 1000$									
	Dimensionless distance, r_D									
	1.1	1.2	1.5	2.0	3.0	5.0	7.0	10.0	20.0	30.0
5.0	934	875	726	543	310	97	26	2	0	0
	940	886	746	568	332	101	24	2	0	0
10.0	945	896	771	614	404	180	77	18	0	0
	949	903	784	631	422	188	77	16	0	0
20.0	953	912	804	668	481	266	148	59	1	0
	956	916	813	681	497	277	153	57	1	0
50.0	961	926	837	723	564	370	253	144	18	1
	963	929	843	731	574	381	378	146	16	0
100.0	967	935	856	755	613	437	326	216	55	11
	967	937	860	760	621	446	334	222	53	10
200.0	969	942	871	780	653	493	390	285	109	39
	970	943	874	784	658	500	397	291	110	38
300.0	971	945	879	793	673	522	424	323	144	65
	972	946	881	796	677	528	430	328	146	64
500.0	973	949	887	807	695	554	462	365	190	102
	974	950	889	810	699	559	468	372	194	104
700.0	974	951	892	815	708	573	484	392	219	129
	975	952	893	818	712	578	490	397	223	132
1000.0	975	953	897	824	721	591	506	417	249	159
	976	954	898	826	724	596	511	422	254	162

19.2 Temperature Distribution in Formations during the Shut-In Period

Knowledge of the temperature distribution around the wellbore as a function of the circulation time, shut-in time, and the radial distance is needed to estimate the electrical resistance of the formation water. This will permit to improve the quantitative interpretation of electric logs. The temperature distribution around a shut-in well is an important factor affecting thickening time of cement, rheological properties, compressive strength development, and set time. For the fluid circulating period an approximate analytical solution was obtained (Eq. (3-74)), which describes with high accuracy the temperature field of formations around a well with a constant bore-face temperature. Using the principle of superposition for the shut-in period we present an approximate analytical solution which describes the temperature distribution in formation surrounding the wellbore during the shut-in period:

$$T_{sD} = \frac{T_s(r,t_s) - T_f}{T_w - T_f} = \frac{Ei\left(-\dfrac{r_D^2}{4(t_D^* + t_{sD})}\right) - Ei\left(-\dfrac{r_D^2}{4t_{sD}}\right)}{Ei\left(-\dfrac{r_D^2}{4t_D^*}\right)},$$

(19-2)

$$t_{sD} = \frac{at_s}{r_w^2},$$

where t_{sD} is the dimensionless shut-in time and t_s is the shut-in time. The values of dimensionless radial temperature of formations calculated after Eq. (19-2) are in good agreement with the results of a numerical solution (Table 19-2).

Table 19-2. Dimensionless shut-in temperature $T_{sD} \cdot 1000$. First line is obtained from Eq. (19-2), second line—numerical solution (Taylor 1978).

t_D	r_D	\multicolumn{8}{c}{Dimensionless shut-in time}							
		5	10	20	30	40	50	70	100
10	1	374	248	151	109	86	71	52	38
		369	247	150	108	85	70	52	-
	2	344	236	147	107	84	70	52	37
		342	235	146	106	84	69	51	-
	3	301	217	140	103	82	68	51	37
		300	216	139	103	81	67	50	-
	5	199	167	120	93	75	63	48	35
		199	166	119	92	75	63	48	-
100	1	569	458	349	290	251	222	181	144
		-	459	352	292	251	223	182	144
	2	545	446	344	287	248	220	180	143
		-	447	346	289	250	221	181	143
	3	509	427	335	281	244	217	178	142
		-	429	337	283	246	218	179	142
	5	418	374	308	263	231	207	172	138
		-	377	310	265	233	208	173	138

A computer program "Shutemp" (Kutasov 1999, pp. 312–313) has been used to calculate the function $T_{sD} = T_{sD}(r_D, t_D, t_{sD})$.

Example

A well was drilled to a depth of 12,490 ft in Webb County, Texas (Venditto and George 1984). The values of the static temperature of formations and circulating temperature at the bottom-hole are: $T_f = 306°F$ and $T_m = T_w = 251°F$. Let us assume that after 50 hours of mud circulation the well was shut-in for 100 hours and after that an electrical log was run (near the bottom-hole); radius of investigation, bit diameter, and the thermal diffusivity of formations are: $r_{inv} = 22$ in., $2_{rw} = 8.75$ in., $a = 0.04$ ft/hr.

The following steps are needed to calculate the radial temperature:

Step 1. Compute the dimensionless circulation time, dimensionless shut-in time, and dimensionless radial distance (dimensionless radius of investigation).

$$t_D = \left(50 \cdot 0.04 \cdot \frac{(2 \cdot 12)^2}{8.75^2}\right) = 15, t_{sD} = \left(100 \cdot 0.04 \cdot \frac{(2 \cdot 12)^2}{8.75^2}\right) = 30, r_D = 22.0 \cdot \frac{2}{8.75} = 5.$$

Step 2. From Eq. (19-2) determine the value of function T_{sD} for $t_D = 15$, $t_{sD} = 30$, and $r_D = 5$. The value of T_{sD} is 0.115 and $T(r_{inv}, t_s) = 0.115 \cdot (251 - 306) + 306 = 300(°F)$.

Step 3. From Eq. (19-2) we can estimate the function T_{sD} at $r = r_w$. The value of T_{sD} is 0.134 and $T(r_w, t_s) = 0.134 \cdot (251 - 306) + 306 = 299(°F)$.

Thus, in this example the radial temperature of formations changes from 299°F at the wellbore face to 300°F at $r_{inv} = 22$ in. The temperatures are close to the static temperature of 306°F (152°C) rather than to the value of the circulating temperature 251°F (122°C). This should be taken into account at interpretation of electric logs.

20

Cylindrical Probe with a Constant Temperature: Determination of the Formation Thermal Conductivity and Contact Thermal Resistance

A new technique has been developed for determination of the formation thermal conductivity and contact thermal resistance. It is assumed that the volumetric heat capacity of formations is known and the instantaneous heat flow rate and time data are available for a cylindrical probe with a constant temperature placed in a wellbore. A semi-theoretical equation is used to approximate the dimensionless heat flow rate. The accuracy of the basic equation and simulated example are shown below.

Due to the similarity in Darcy and Fourier laws the same differential diffusivity equation describes the transient flow of incompressible fluid in porous medium and heat conduction in solids. As a result, a correspondence exists between the following parameters: volumetric flow rate, pressure gradient, mobility (formation permeability and viscosity ratio), hydraulic diffusivity coefficient and heat flow rate, temperature gradient, thermal conductivity and thermal diffusivity. Thus, the same analytical solutions of the diffusivity equation (at corresponding initial and boundary conditions) can be utilized for determination of the above-mentioned parameters. Earlier we suggested a semi-theoretical equation to approximate the dimensionless heat flow rate from an infinite cylindrical source with a constant bore-face temperature (Kutasov 1987). This equation was used to process data of pressure and flow well tests and to develop a technique for determining the formation permeability and skin factor (Kutasov 1998; Kutasov and Kagan 2000, 2001).

The objective of this chapter is to suggest a similar technique for *in situ* evaluation of the values of formation thermal conductivity and thermal resistance of the borehole (expressed through the skin factor). For this reason we did not conduct a review and analysis of relevant publications here. We will only mention the interesting Temperature Recovery Method (TRM) (Günzel and Wilhelm 2000). The TRM can be used to determine *in situ* thermal properties of formations and the thermal resistance of the borehole wall. The thermal disturbance is modelled by a perfectly conducting cylindrical source of constant thermal power per unit of length q in a borehole with volumetric heat capacity C_{vb} and a thermal resistance R drilled in homogeneous isotropic rock with volumetric heat capacity C_v and thermal conductivity λ. At a known T_f the solution for this model (Carslaw and Jaeger 1959) is function of the above mentioned parameters and time. For a reliable determination of the thermal resistance of the wall, at least six temperature measurements are required within the first out of recovery after generation of a temperature disturbance. The temperature data are evaluated using an inversion algorithm. This algorithm provides an optimum value and an error estimate for each parameter (Günzel and Wilhelm 2000). It should be noted that TRM cannot be applied in the cased boreholes.

Below we will consider a long cylindrical electrical heater (with a large length/diameter ratio). Calculations conducted by Mufti (1971) revealed that for practical purposes a cylinder whose length is 5 times or more its diameter could be treated as an infinite cylinder. In this case the heater can be considered as an infinite cylindrical source of heat. For this case the temperature field in and around the borehole is a function of time, radial distance, thermal diffusivity of formations, and borehole thermal resistance. An effective radius concept is introduced to evaluate the effect of the contact thermal resistance on the heat flow rate into formation.

20.1 Effective Radius of the Heater

To take into account the effect of probe's casing and the contact thermal resistance on the heat flow rate we will use an effective radius concept (Uraiet and Raghavan 1980). This approach is widely used in transient pressure and flow well testing (Earlougher 1977) to evaluate the effect of formation damage (improvement) around the borehole on the pressure at the borehole's wall. Firstly, we introduce skin factor (s)—a parameter which allows us to quantitatively determine the effect of the well's thermal resistance on the heat flow rate. In our case

$$s = \left(\frac{\lambda}{\lambda_{ef}} - 1 \right) \ln \frac{r_w}{r_h}, \tag{20-1}$$

where r_w is the well radius, r_h is the radius of the heater (without casing), λ is the thermal conductivity of formations (around the borehole), λ_{ef} is the effective thermal conductivity of the $r_w - r_h$ annulus, and r_{ha} is the effective radius of the heater.

For an open (uncased) borehole the $r_w - r_h$ annulus is filled with the drilling fluid (or air) and mud cake–a plastic like coating of the borehole resulting from the solids in the drilling fluid adhering and building up on the wall of the hole. The $r_w - r_h$ ring in a cased borehole is composed of drilling fluid, steel, and cement.

It is more convenient to express the skin factor through the apparent (effective) heater radius (Earlougher 1977)

$$r_{ha} = r_h \exp(-s). \tag{20-2}$$

20.2 Dimensionless Flow Rate

Let us assume that the probe (at $r = r_h$) is maintained at a constant temperature of T_h and the initial (undisturbed) temperature of formations is T_f. In this case and the effective wellbore radius concept can be employed (Earlougher 1977). In this case the relationship between the heat flow rate per unit of depth (q) and time is:

$$q = 2\pi\lambda\left(T_h - T_f\right)q_D,$$ (20-3)

$$t_D = \frac{\lambda t}{\rho c_p r_{ha}^2},$$ (20-4)

$$r_{wa} = r_w e^{-s},$$

where q_D is dimensionless heat flow rate, ρc_p is volumetric heat capacity of formations, and t_D is the dimensionless time. We will assume that the volumetric heat capacity of formations is known. Analytical expressions for the function $q_D = f(t_D)$ are available only for asymptotic cases or for large values of t_D. The dimensionless flow rate was first calculated and presented in a tabulated form by Jacob and Lohman (1952). Sengul (1983) computed values of q_D for a wider range of t_D and with more table entries. We have found (Kutasov 1987) that for any values of dimensionless production time a semi-theoretical Eq. (20-5) can be used to forecast the dimensionless heat flow rate:

$$q_D = \frac{1}{\ln\left(1 + D\sqrt{t_D}\right)},$$ (20-5)

$$D = d + \frac{1}{\sqrt{t_D} + b}, \qquad d = \frac{\pi}{2}, \qquad b = \frac{2}{2\sqrt{\pi} - \pi}.$$ (20-6)

In Table 20-1 the values of q_D calculated after Eqs. (20-4) and (20-5) and the results of a numerical solution for q_D^* (Sengul 1983) are compared. The agreement between values of q_D and q_D^* calculated by these two methods is seen to be good.

Table 20-1. Comparison of values of dimensionless heat flow rate for cylindrical source with a constant temperature: q_D^* is the numerical solution (Sengul 1983); q_D is after application of Eq. (20-5).

t_D	q_D^*	q_D	$\Delta q/q^* \cdot 100\%$
0.0001	56.918	56.930	0.02
0.001	18.337	18.350	0.07
0.01	6.1289	6.1410	0.20
0.1	2.2488	2.2596	0.48
1	0.98377	0.99260	0.90
10	0.53392	0.54068	1.27
100	0.34556	0.35025	1.36
1000	0.25096	0.25366	1.08
10000	0.19593	0.19727	0.69

20.2.1 Working formula

Let us now assume that the heater starts to operate at time $t = 0$ and at least two values of the heat flow rate per unit of length (q_1 and q_2) for $t = t_1$ and $t = t_2$ are available. Then,

$$q_1 = 2\pi\lambda\left(T_h - T_f\right)q_D\left(t_{D1}\right), \quad t_{D1} = \frac{\lambda t_1}{\rho c_p r_{ha}^2}, \tag{20-7}$$

$$q_2 = 2\pi\lambda\left(T_h - T_f\right)q_D\left(t_{D1}\beta\right), \quad t_{D2} = \frac{\lambda t_2}{\rho c_p r_{ha}^2} = t_{D1} \cdot \beta, \quad \beta = \frac{t_2}{t_1}. \tag{20-8}$$

Combining the last two equations we obtain

$$\gamma = \frac{q_1}{q_2} = \frac{q_D\left(t_{D1}\right)}{q_D\left(t_{D1}\,\beta\right)}. \tag{20-9}$$

Let us now assume that the absolute accuracy of the ratio γ is ε, and then an equation for calculating the value of t_{D1} is

$$\gamma - \frac{q_D\left(t_{D1}\right)}{q_D\left(t_{D1}\,\beta\right)} = \varepsilon. \tag{20-10}$$

The formation thermal conductivity is determined from Eq. (20-7). The thermal diffusivity of formations is estimated from the relationship

$$a = \frac{\lambda}{\rho c_p}, \tag{20-11}$$

where ρ is the formation's density and c_p is the specific heat. The apparent (effective) heater radius and skin factor are calculated from Eqs. (20-2) and (20-4):

$$r_{ha} = \sqrt{\frac{\lambda t}{\rho c_p t_{D1}}}, \quad s = -\ln\frac{r_{ha}}{r_w}. \tag{20-12}$$

And, finally, the values of λ_{ef} and R are evaluated from Eq. (20-1):

$$\lambda_{ef} = \frac{\lambda \ln\frac{r_w}{r_h}}{s + \ln\frac{r_w}{r_h}}, \quad R = \frac{1}{\lambda_{ef}}, \quad r_w \neq r_h. \tag{20-13}$$

20.2.2 Simulated example

A metallic electrical heater was placed into a vertical open (uncased) borehole. The heater operated for 10 hours and the transient heat flow rate was recorded (Table 20-2). The well radius $r_w = 0.10$ m, the radius of the probe $r_h = 0.08$ m. The $r_w - r_h$ annulus consists of mud cake and drilling fluid. We assumed that the effective thermal conductivity of the $r_w - r_h$ annulus is $\lambda_{ef} = 0.9741$ W/m°C and thermal contact resistance

$R = 1/\lambda_{ef} = 1.027$ m°C/ W. The heater temperature (T_h) is 60°C and the initial formation temperature (T_f) is 40°C. The heater ($T_h - T_f = 20$°C) operated for 10 hours and the transient heat flow rate was recorded. The formation is sandstone with $\rho = 2300$ kg/ m³, $\lambda = 2.000$ W/m°C, and $c = 783$ J/kg °C. Using the table (Sengul 1983) of $q_D{}^* = f(t_D)$ we generated data for this simulated example. The input data were chosen to allow to avoid interpolation of $q_D{}^*$ values. The results of calculations after Eqs. (20-1)–(20-13) are presented in Table 20-2. This example shows that the basic Eq. (20-5) can be used to compute the thermal conductivity of formations and contact thermal resistance. Indeed, the assumed and calculated values of λ and R are in a very good agreement.

Table 20-2. Comparison of assumed and calculated values of formation thermal conductivity and thermal contact resistance. Assumed: $\lambda = 2.000$ W/m·°C, $R = 1.027$ m·°C/ W.

t_1, hrs	t_2, hrs	q_1, W/m	q_2, W/m	λ, W/m·°C	R, m·°C/W
1	2	247.25	201.21	1.956	1.027
1	3	247.25	180.00	1.956	1.027
1	5	247.25	157.88	1.956	1.027
2	4	201.21	166.98	1.957	1.027
2	6	201.21	151.02	1.957	1.027
2	8	201.21	141.14	1.957	1.028
3	6	180.00	151.02	1.957	1.028
3	8	180.00	141.14	1.957	1.028
3	10	180.00	134.19	1.958	1.029
4	8	166.98	141.14	1.957	1.028
4	10	166.98	134.19	1.958	1.029
5	10	157.88	134.19	1.958	1.029

A method of determination *in situ* of formation thermal conductivity and thermal resistance of the borehole is developed. This method is based on a semi-analytical equation, which approximates the heat flow rate from an infinitely long cylindrical source with a constant bore-face temperature. This equation is valid for any values of dimensionless time. To verify applicability of the suggested method field experiments in open and cased boreholes are needed.

The Reader cannot assume that for the first six runs (Table 20-2) we assumed a perfect contact between the probe and formations ($\lambda_4 = \lambda_a$). In this case one can observe maximal values of the radial heat flow rate.

21

Well Temperature Testing: An Extension of the Slider's Method

A new technique has been developed for determination of the formation thermal conductivity, skin factor, and contact thermal resistance for boreholes where the temperature recovery process after drilling operations is not completed. Slider suggested a technique for analyzing transient pressure tests when conditions are not constant. We extend the Slider's method for transient temperature well tests (Kutasov and Eppelbaum 2007). It is assumed that the volumetric heat capacity of formations is known, and the instantaneous heater's wall temperature and time data are available for a cylindrical probe with a constant heat flow rate placed in a borehole. A semi-analytical equation is used to approximate the dimensionless wall temperature of the heater. A simulated example is presented to demonstrate the data processing procedure.

The similar analytical form of Darcy and Fourier laws, which describe the transient flow of an incompressible fluid in a porous medium and the heat conduction in solids, respectively, leads to a correspondence between the thermal and hydraulic parameters. It also important to show that the product of porosity and total compressibility can be obtained from multi-well tests (interference testing). Fortunately, the analogous thermal parameter—the product of density and specific heat—varies within narrow limits and can be determined from cuttings (Kappelmeyer and Haenel 1974; Somerton 1992). Below we will also show that an additional parameter, the well's contact thermal resistance, can be expressed through the skin factor and the formation thermal conductivity. Thus, the same analytical solutions of the diffusion equation (at corresponding initial and boundary conditions) can be utilized for the determination of the hydraulic and thermal parameters.

However, this approach can be used only for large dimensionless times (when the solution of the diffusion equation can be expressed by an exponential integral). Generally the mathematical model of pressure well tests is based on the assumption that the borehole is an infinitely long linear source with a constant flow rate in an infinite, homogeneous reservoir. In this case, the well-known solution of the diffusion equation is an exponential integral. In thermal measurements the borehole (or the cylindrical heater) cannot be considered as an infinite long linear source of heat. This is due to low thermal diffusivity of rocks in comparison with the hydraulic diffusivity and corresponding low values of dimensionless time. Kutasov (2003) showed that the convergence of solutions of the diffusion equation for cylindrical and linear sources occurred at dimensionless time of about 1000. A semi-theoretical equation was suggested earlier (Kutasov 1987) to approximate the dimensionless heat flow rate from a cylindrical source with constant bore-face temperature. This equation was used to process pressure and temperature data from well tests and to develop a technique for determining the formation permeability, skin factor, thermal conductivity and thermal resistance of the borehole (Kutasov 1998; Kutasov and Kagan 2003a, 2003b; Kutasov and Eppelbaum 2005b, 2007).

Eppelbaum and Kutasov (2006a, 2013) proposed a new technique, based on semi-analytical equation for the dimensionless temperature at the wall of an infinite long cylindrical source with a constant heat flow rate (Kutasov 2003). This method enables to determine the formation thermal conductivity, the contact thermal resistance and formation temperature. The utilization of this method requires that (prior to the test) the thermal recovery is practically completed. However, the drilling process greatly alters the temperature of rocks immediately surrounding the well. The temperature change is affected by the duration of drilling fluid circulation, the temperature difference between the reservoir and the drilling fluid, the well radius, the thermal diffusivity of the reservoir and the drilling technique. Thus, the determination of formation temperature (with a specified absolute accuracy) at any depth requires a lengthy period of shut-in time.

The objective of this chapter is to suggest a similar technique for *in situ* evaluation of thermal conductivity and thermal resistance (expressed as skin factor) for boreholes where thermal recovery is not completed (short shut-in periods). Here we must note that Sass et al. (1981) suggested *in situ* determination of heat flow in boreholes by the use of cable power supplying. We will consider a long cylindrical electrical heater (cylindrical heater is an analogue of radius of borehole with skin in pressure well testing) with a large length/diameter ratio. Mufti (1971) demonstrated that for practical purposes a cylindrical heater whose length is 5 times or more its diameter could be treated as an infinite cylindrical source of heat. Thus, the temperature field in and around the borehole is a function of time, radial distance, rock thermal diffusivity, and borehole thermal resistance. The "effective radius" concept must be introduced to evaluate the effect of the contact thermal resistance on the heat flowing into the formation (for applying the proposed procedure is necessary to select a comparatively homogeneous interval of geological strata). For validating the proposed method we use an "exact" solution (numerical) and generate synthetic data for a test. Then the semi-analytical equation was used to process the results and to compare the obtained and assumed input parameters. The final step in validation of the proposed technique is to conduct a number of field tests and to compare the results of these tests with those obtained by means of other methods. A simulated example is presented to demonstrate the data processing procedure for determination of formation thermal conductivity, skin factor, and contact thermal resistance.

21.1 Slider's Method

The Slider's method is a technique for analyzing transient pressure (p) tests (e.g., Earlougher 1977, pp. 27–29). Figure 21-1 schematically illustrates the shut-in pressure declining (solid line) before a drawdown test's start at time t_1. The dashed line represents the expected pressure behavior with time. If a constant flow rate starts at time t_1, pressure decreases as shown by the solid line. The analysis of such drawdown behavior requires: (1) the correct extrapolation of the shut-in pressure, (2) the estimation of the difference between observed pressure and the extrapolated pressure ($\Delta p_{\Delta t}$), (3) plotting $\Delta p_{\Delta t}$ versus log Δt.

For extending this method to the temperature analysis it is necessary to estimate the rate (U) of temperature change at $t = t_1$ and to determine the values of the temperature decline ("correct extrapolation" like in Figure 21-1). Corrections finally should be introduced to the observed (while testing) well temperatures.

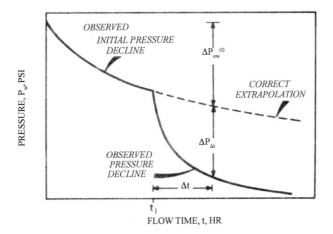

Figure 21-1. Drawdown testing in a developed reservoir, definition of terms (Earlougher 1977).

21.2 Effective Radius of the Heater

We will use the effective radius concept to take into account the effect of probe casing and the contact thermal resistance on the heat flow rate. This approach is widely used in transient pressure and flow well testing (Earlougher 1977) to evaluate the effect of formation permeability changing around the borehole on the pressure at the borehole's wall. Firstly, we introduce skin factor (s)—a parameter which allows quantitatively to determine the effect of the well thermal resistance on the heat flow rate. In our case

$$s = \left(\frac{\lambda}{\lambda_{ef}} - 1 \right) \ln \frac{r_w}{r_h} , \qquad r_w \neq r_h , \tag{21-1}$$

where r_w is the well radius, r_h is the radius of the heater (without casing), λ is the rock thermal conductivity and λ_{ef} is the effective thermal conductivity of the $r_w - r_h$ annulus (Kutasov and Eppelbaum 2007) (Figure 21-2).

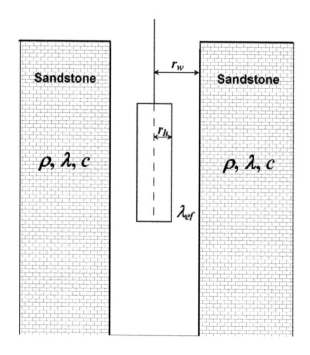

Figure 21-2. Formation-wellbore-probe system.

It is more convenient to express the skin factor through the apparent (effective) heater radius (Earlougher 1977)

$$r_{ha} = r_h \exp(-s), \tag{21-2}$$

where r_{ha} is the effective radius of the heater.

For an open (uncased) borehole the $r_w - r_h$ annulus is filled with the drilling fluid (or air), plastic like mud cake coating the borehole. This results from the solids in the drilling fluid adhering to the wall of the hole. The $r_w - r_h$ ring in a cased borehole is composed of drilling fluid, steel, and cement. The skin factor can be estimated from a temperature drawdown test. The thermal contact resistance $R = \dfrac{1}{\lambda e_f}$ is easily calculated from Eq. (21-1):

$$\lambda_{ef} = \frac{\lambda \ln \dfrac{r_w}{r_h}}{s + \ln \dfrac{r_w}{r_h}}, \qquad R = \frac{s + \ln \dfrac{r_w}{r_h}}{\lambda \ln \dfrac{r_w}{r_h}}. \tag{21-3}$$

We should mention that the skin factor was introduced in petroleum engineering by Hawkins (1956) to account for the pressure drop in the zone (around the wellbore) of altered permeability. The skin factor is a composite parameter and it takes into account permeability of different layers by introducing the effective (equivalent) permeability. The reliability of estimation of the skin factor depends only on the quality of the field pressure and flow data (test design, type of instrumentation, data processing technique, an adequate physical and mathematical model). Similarly, for a temperature test, the skin factor takes into account the thermophysical properties of the materials (e.g., drilling fluids, steel, cement, etc.) through the effective thermal conductivity of the wellbore-heater.

21.3 Rate of Temperature Decline

We selected long term temperature data from four two wells to demonstrate the application of the Slider's method for analyzing results of temperature test in deep wells (Tables 21-1–21-3 and Figure 21-3).

Table 21-1. Well data and references.

Borehole	Well 192	PBF
Site name	Kugpik D-13	Put River N-1
Location	Lat: 68 52.8 N	Lat: 70 19 07 N
	Long: 135 18.2 W	Long: 148 54 35 W
Hole depth, m	3689	763
Drilling time (days)	188	44
Number of logs	7	9
Shut-in period (days)	35-2835	5-1071
Reference	Taylor et al. 1982	Clow and Lachenbruch 1998

Table 21-2. Calculated rates of temperature decline (U), observed temperatures, values of T^* and formation temperature (T_f), well 192, Kugpik D-13.

z, m	T_1, °C $t_{s1} = 35$ d	T_2, °C $t_{s2} = 128$ d	T_3, °C $t_{s3} = 320$ d	T^*, °C	$-U \cdot 10^{-4}$, °C/hr at t_{s2}	T_f, °C
121.6	8.43	5.34	3.42	2.43	6.28	1.37
152.4	9.67	6.58	4.38	3.68	6.27	1.94
182.9	10.67	7.66	5.51	4.85	6.10	3.13
213.4	11.45	8.63	6.80	6.00	5.71	4.83
243.8	11.89	9.48	7.82	7.24	4.87	6.01
274.6	11.66	9.84	8.34	8.16	3.68	6.61
305.1	12.41	10.48	9.07	8.70	3.89	7.51
335.6	13.12	11.33	9.80	9.69	3.61	8.03
366.1	13.68	11.99	10.55	10.44	3.40	8.89
396.5	14.27	12.55	11.25	10.98	3.46	9.81
427.3	14.79	13.23	11.94	11.81	3.13	10.47
457.8	15.41	13.83	12.69	12.40	3.17	11.45
488.6	16.43	14.99	13.65	13.69	2.88	12.07
519.1	16.78	15.54	14.43	14.43	2.48	13.14

Table 21-3. Observed shut-in temperatures, well PUT River, Alaska N-1.

z, m	Shut-in time, days				
	5	22	34	48	1071
30.48	−.400	−2.686	−4.793	−6.252	−9.167
45.73	−.300	−2.093	−4.507	−6.012	−9.052
60.96	−.250	−2.941	−4.911	−6.148	−8.957
91.45	−.300	−1.633	−4.101	−5.646	−8.771
152.40	−.030	−.976	−1.852	−3.173	−8.124
304.81	.740	−.379	−.506	−.682	−5.462
335.28	.910	−.325	−.451	−.577	−4.935
396.24	1.230	−.354	−.505	−.644	−4.039
579.13	2.540	−.010	−.236	−.309	−.778
609.61	2.600	3.491	1.918	1.212	−.195
640.08	6.830	5.275	4.047	3.254	.761
670.56	9.350	5.948	4.749	4.002	1.664
701.04	9.910	7.251	6.022	5.256	2.885

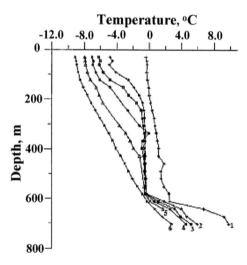

Figure 21-3. Temperature profiles in the Put River N-1 well. The shut-in times for curves 1, 2, 3, 4, 5, and 6 are 5, 34, 48, 66, 117, and 1071 days respectively (Clow and Lachenbruch 1998).

Well 192, Kugpik D-13

For the 121.6–519.1 m section of this well (Table 21-2) we assumed that the shut-in the temperature recovery for short shut-in times could be approximated by the Horner plot

$$T_s = T* + B \ln \frac{t_s + t_c}{t_s}, \tag{21-4}$$

where T^* is the temperature trend extrapolated to infinite shut-in time, T_s is the shut-in temperature, t_s is the shut-in time, B is the coefficient, and t_c is the thermal "disturbance" period for a given depth.

It is a reasonable assumption that the value of t_c is a linear function of the depth (z):

$$t_c = t_{tot}\left(1 - \frac{z}{H_t}\right), \qquad (21\text{-}5)$$

where z is the depth, t_{tot} is the total drilling time, and H_t is the total well depth.

The values of T^* were computed by using values of shut-in temperature at $t_s = t_{s1}$ and $t_s = t_{s2}$ (Table 21-2). To show that the Horner plot method cannot be applied in this case for determining formation temperatures we used the Three Point Method (Kutasov and Eppelbaum 2003, 2010) to estimate the value of T_f (undisturbed formation temperature)—please compare columns 5 and 7 in Table 21-2. From Eq. (21-4) the rate of the temperature change is

$$U = \frac{dT_s}{dt_s} = B\left(\frac{1}{t_s + t_c} - \frac{1}{t_s}\right). \qquad (21\text{-}6)$$

The calculated values of U are presented in Table 21-2. Let now assume that after 128 days of shut-in time we carry out a temperature drawdown test, the absolute accuracy of field temperature measurements is 0.01°C and the duration of the test is 10 hours. Then, for the depth of 121.6 m (the highest absolute value of U) the maximum correction to the observed temperature is $\Delta T_s = 0.000628 \cdot 10 = 0.0063$ (°C). It is evident that in this case we can consider that the initial formation temperature is 5.34°C (Table 21-2).

Well PUT River N-1

We used values of transient temperature for two depths 670.56 and 701.04 m for shut-in time of 1071 days (Table 21-3) to estimate the position of the 0°C isotherm. The linear extrapolation gives that the base of permafrost is located at about 629 m. Let us now assume that at time $t = t_x$ the phase transition (water-ice) in formation at a selected depth is completed, i.e., the thermally disturbed formation has frozen ends. In this case at $t > t_x$ the cooling process is similar to that of temperature recovery in sections of the well below the permafrost base. It is well known (Tsytovich 1975) that the freezing of the water occurs in some temperature interval below 0°C (Figure 21-4).

In practice, however, t_x cannot be determined. This can be realized only by conducting long-term continuous repetitive temperature observations in deep wells. For the permafrost interval we assumed that at $t_s > t_x$ and for short shut-in times the shut-in temperatures can be approximated by a simple equation

$$T_s = T_{ep} + C\ln\frac{t_s}{t_x}, \qquad (21\text{-}7)$$

where C is a coefficient.

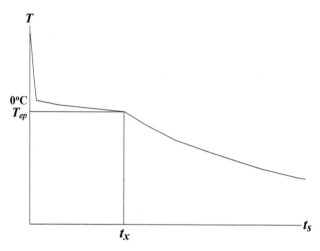

Figure 21-4. Shut-in temperatures at a given depth—schematic curve.

It was assumed that $T_{ep} = 0°C$ and the values of C and t_x were estimated by using values of shut-in temperature at $t_s = t_{s1}$ and $t_s = t_{s2}$. Then the rate of the temperature decline is

$$U = \frac{dT_s}{dt_s} = C\frac{1}{t_s}.$$ (21-8)

For the depths $z > 640$ m Eq. (21-6) was used to estimate the values of U. The rates of the temperature decline are presented in Table 21-4.

Table 21-4. Calculated rates of temperature decline (U), Put River N-1 Well.

z, m	t_{s1}, days	t_{s2}, days	T_1, °C	T_2, °C	$-U \cdot 10^{-3}$, °C/hr at t_{s2}
30.48	22	34	−2.686	−4.793	5.932
45.72	22	34	−2.093	−4.507	6.796
60.96	22	34	−2.941	−4.911	5.546
91.44	22	34	−1.633	−4.101	6.948
30.48	34	48	−4.793	−6.252	3.673
45.72	34	48	−4.507	−6.012	3.788
60.96	34	48	−4.911	−6.148	3.114
91.44	34	48	−4.101	−5.646	3.889
640.08	5	22	6.830	5.275	1.189
670.56	5	22	9.350	5.948	2.470
701.04	5	22	9.910	7.251	1.810

21.4 Well as a Cylindrical Source

A long cylindrical electrical heater (large length/diameter ratio) with a constant heat flux is often used in the laboratory for determination of the thermal conductivity of samples of rocks. In this case the transient temperature T_w is a function of time, thermal conductivity, and volumetric heat capacity of formations. Analytical expression for T_w is available only for large values of the dimensionless time t_D expressed by

$$t_D = \frac{\chi t}{r_w^2} = \frac{\lambda t}{c_p \rho r_w^2},$$ (21-9)

$$r_{wa} = r_w e^{-s}.$$

where χ is the thermal diffusivity of formations, t is the time, r_w is the well radius, λ is the thermal conductivity, c_p is the specific heat at constant pressure, and ρ is the density.

To determine the temperature T_w it is necessary to solve the diffusion equation under the following boundary and initial conditions:

$$T(t = 0, r) = T_f, \qquad r_w \leq r < \infty, \qquad t > 0,$$ (21-10)

$$\left(r \frac{\partial T}{\partial r} \right)_{r_w} = \frac{q}{2\pi\lambda}, \qquad t > 0,$$ (21-11)

$$T(t, r \to \infty) \to T_f, \qquad t > 0,$$ (21-12)

where T_i is the initial (formation) temperature, r is the radial distance, and q is the heat flow rate per unit of length.

It is well known that in this case the diffusion equation has a solution in complex integral form (Van Everdingen and Hurst 1949; Carslaw and Jaeger 1959). Chatas (Lee 1982, pp. 106–107) tabulated this integral for $r = r_w$ over a wide range of t_D values.

A semi-theoretical equation of the wall dimensionless temperature (T_D) for a cylindrical source with a constant heat flow rate is (Kutasov 2003):

$$T_D(t_D) = \ln\left[1 + \left(c - \frac{1}{a + \sqrt{t_D}} \right) \sqrt{t_D} \right],$$ (21-13)

$$a = 2.7010505, \qquad c = 1.4986055,$$

$$T_D(t_D) = \frac{2\pi\lambda(T_w - T_f)}{q}.$$ (21-14)

Kutasov (2003) compared the values of T_D calculated from Eq. (21-13) and results of a numerical solution ("Exact" solution) by Chatas (Lee 1982, pp. 106–107). The agreement between values of T_D calculated by these two methods is very good. For this reason the principle of superposition can be used without any limitations.

21.4.1 Temperature drawdown well test

Let us assume that the initial formation temperature (prior to the test) is known. At least two measurements of wall temperature (at time $t = t_1$ and $t = t_2$) are needed to calculate the formation thermal conductivity, skin factor, and thermal contact resistance. As is customary in petroleum engineering the effect of the skin factor can be expressed by introducing the dimensionless time t_{Da} based on the apparent well radius r_{ha}.

Let

$$m = \frac{q}{2\pi\lambda}, \qquad t_{Da} = \frac{\lambda t}{\rho c_p r_{ha}^2}, \tag{21-15}$$

$$F(t_{Da}) = \ln\left[1 + \left(c - \frac{1}{a + \sqrt{t_{Da}}}\right)\sqrt{t_{Da}}\right]. \tag{21-16}$$

Then

$$T_h = T_i + mF(t_{Da}) \tag{21-17}$$

and

$$\gamma = \frac{T_{h1} - T_f}{T_{h2} - T_f} = \frac{F(t_{Da1})}{F(t_{Da2})} = \psi(t_{Da1}), \tag{21-18}$$

$$t_{Da1} = \frac{\lambda t_1}{\rho c_p r_{ha}^2}, \qquad t_{Da2} = t_{Da1}\beta_1, \qquad \beta_1 = \frac{t_2}{t_1}. \tag{21-19}$$

If we assume that the absolute accuracy of the ratio γ is ε, then solving the following equation we calculate the value of t_{Da1}:

$$\gamma - \psi(t_{Da1}) = \varepsilon \tag{21-20}$$

and from equation

$$T_{h_1} = T_i + mF(t_{Da_1}) \tag{21-21}$$

we can calculate the value of m. Then the formation thermal conductivity can be determined

$$\lambda = \frac{q}{2\pi m} \tag{21-22}$$

and, finally the skin factor and the thermal contact resistance per unit of length can be estimated from Eq. (21-3). Let us assume that we plan to conduct two drawdown temperature tests in the Well PUT River N-1 (Alaska) at the depths of 91.44 and 670.56 m after 34 and 22 days of shut-in respectively. In the first case the initial temperature is –4.101°C (Table 21-3) but the observed temperatures (T_{obs}) should be corrected

$$T_{corr} = T_{obs} - U. \tag{21-23}$$

For this test, to avoid thawing of frozen formation cooling of formations is desirable (using a cylindrical heat sink). For the second test the initial temperature is 5.948°C (Table 21-3) and the observed temperatures should be corrected (Eq. (21-23)) and an electrical heater (probe) can be used (see the simulated example below).

21.4.2 Simulated example

A metallic electrical heater is placed into Well Put River N-1 (uncased) at the depth 670.56 m. The test is conducted after 22 days of shut-in and the rate of temperature decline is $U = -2.470 \cdot 10^{-3}$°C/hr (Table 21-4). The heater generates a heat flow into the formation of 80.0 W/m and operated for 10 hours. The transient heater's wall temperature was recorded (Table 21-5). The well radius $r_w = 0.10$ m, the radius of the probe $r_h = 0.08$ m. The $r_w - r_h$ annulus consists of mud cake and drilling fluid. We assume that the effective thermal conductivity of the $r_w - r_h$ annulus is $\lambda_{ef} = 0.9741$ W/m·°C and thermal contact resistance is $R = 1/\lambda_{ef} = 1.027$ m·°C/W. The initial formation temperature (T_i) is 5.948°C. The formation is sandstone with $\rho = 2300$ kg/m³, $\lambda = 2.000$ W/m·°C, and $c = 783$ J/kg °C. Using the table of Chatas (Lee 1982, pp. 106–107) of $T_D = f(t_D)$ we generated data for this simulated example. The input data were chosen to avoid interpolation of T_D values. The results after Eqs. (21-3) and (21-16)–(21-22) are presented in Table 21-5. The example shows that the basic Eq. (21-13) can be used to compute the thermal conductivity of formations and contact thermal resistance. Indeed, the assumed and calculated values of λ and R are in a good agreement.

Table 21-5. Comparison of assumed and calculated values of formation thermal conductivity and thermal contact resistance. Assumed: $\lambda = 2.000$ W/m °C, $R = 1.027$ m °C/W.

t_1, hrs	t_2, hrs	\multicolumn{4}{c	}{$U = -0.00247$°C/hr}	\multicolumn{2}{c}{$U = 0$}
		T_{h1}, °C	T_{h2}, °C	λ, W/m·°C	R, m·°C/W	λ, W/m·°C	R, m·°C/W
1	2	11.051	12.433	2.061	1.055	2.051	1.052
1	3	11.051	13.368	1.994	1.025	1.985	1.022
1	5	11.051	14.610	2.002	1.029	1.991	1.025
2	4	12.433	14.055	1.952	0.990	1.940	0.985
2	6	12.433	15.076	1.976	1.005	1.963	0.999
2	8	12.433	15.832	1.986	1.011	1.971	1.005
3	6	13.368	15.076	2.015	1.040	1.999	1.032
3	8	13.368	15.832	2.016	1.041	1.998	1.032
3	10	13.368	16.433	2.019	1.042	2.000	1.033
4	8	14.055	15.832	2.015	1.040	1.997	1.030
4	10	14.055	16.433	2.018	1.042	1.999	1.032
5	10	14.610	16.433	2.021	1.045	1.999	1.032

Thus a new method of determination *in situ* thermal conductivity and thermal resistance in a borehole, where the temperature recovery is not completed (after drilling operations), is proposed. This method is based on a semi-analytical equation, which approximates the dimensionless wall temperature of an infinitely long cylindrical probe with a constant heat flow rate placed into a borehole. It is shown that Slider's method (used in petroleum engineering) can be utilized to analyze results of temperature well tests. Two temperature logs recorded at short shut-in times are required to use the suggested method.

22

Cementing of Casing in Hydrocarbon Wells: The Optimal Time Lapse to Conduct a Temperature Log

Some cases of cementing of casing in hydrocarbon wells are considered in Kutasov and Eppelbaum (2012a, 2013a, 2013b, 2014).

Below we will consider a method of downhole temperature prediction while cementing of casing. It will be shown that for deep and hot wells the heat generation during cement hydration may cause a substantial temperature increase in the annulus. This factor must be taken into account in cement slurry design. Temperature and pressure are two basic influences on the downhole performance of cement slurries. Temperature has the more pronounced influence. The downhole temperature controls the pace of chemical reactions during cement hydration resulting in cement setting and strength development. The shut-in temperature affects how long the slurry will pump and how well it develops the strength to support the pipe. As the formation temperature increases, the cement slurry hydrates and sets faster and develops strength more rapidly. Cement slurries must be designed with sufficient pumping time to provide safe placement in the well. At the same time the cement slurry cannot be overly retarded as this will prevent the development of satisfactory compressive strength. The thickening time of cement is the time that the slurry remains pumpable under set conditions. While retarders can extend thickening times, the thickening time for a given concentration of retarder is still very sensitive to changes in temperature. Slurries designed for erroneously high circulating temperatures can have unacceptably long setting times at lower temperatures.

A new technique has been developed to estimate the temperature increase during cement hydration. The method presented in this chapter will enable to select the optimal time lapse between cement placement and temperature survey. A semi-analytical equation was earlier suggested which describes the transient temperature at the borehole's wall, while the radial heat flow rate (into formations) is a quadratic function of time. Only field or laboratory heat production rate—time data are needed to calculate the transient values of the temperature increase. Two field examples of cement hydration when retarders were used are presented. Assessment of the temperature development during hydration is necessary to determine how fast the cement will reach an acceptable compressive strength before the casing can be released. Therefore, for deep wells heat generation during cement hydration has to be taken into account at cement slurry design. The experimental data show that the maximum value of heat generation occurs during the first 5 to 24 hours. During this period the maximum temperature increase (ΔT_{max}) can be observed in the annulus. In order to evaluate the temperature increase during cement hydration it is necessary to approximate the heat production versus time curve by some analytical function. It was found that a quadratic equation can be used for a short interval of time to approximate the rate of heat generation (q) per unit of length as a function of time. Temperature surveys following the cementing operation are used for locating the top of the cement column behind the casing. Field experience shows that in some cases the temperature anomalies caused by the heat of cement hydration can be very substantial. However, even in such cases it is very important to predict the temperature increase during the cement setting. This will enable to determine the optimal time lapse between cementing and temperature survey. Below we present a semi-analytical formula which will allow one to estimate the temperature increase versus setting time (Kutasov 2007). This formula describes the transient temperature at the cylinder's wall (T_y), while at the surface of the cylinder the radial heat flow rate (into formations) is a quadratic function of time.

For comprehensive design and evaluation of cementing operations a major oilfield service company developed a temperature simulator (Romero and Loizzo 2000). The model incorporates a flow and thermal model, and accommodates complex wellbore geometries. During cement hydration, the thermal diffusivity equation is solved across the casing, cement and formation, and the radial temperature profile in a well configuration is determined. For the study of temperature development in cemented annuli, the degree of hydration is considered as a fundamental parameter. It is defined as the cement fraction that has reacted, and can be calculated practically as the fraction of the heat of hydration has been released:

$$\xi(t) = \frac{Q(t)}{Q_{total}} = \frac{1}{Q_{total}} \int_0^t q(u)\, du,$$

where q is the heat production rate, and Q_{total} is the total heat of hydration.

For cement slurry, the total heat Q_{total} can be calculated based on the slurry composition. A new hydration model was developed, and yields the heat production rate q which is related to the actual degree of hydration and actual temperature.

22.1 Temperature Increase at Cement Hydration

When cement is mixed with water, an exothermic reaction occurs and a significant amount of heat is produced. This amount of heat depends mainly on the fineness and chemical composition of the cement, additives, and ambient temperature.

Temperature surveys following the cementing operation are used for locating the top of the cement column behind the casing. As mentioned, field experience shows that in some cases the temperature anomalies caused by the heat of cement hydration can be very substantial. As in any exothermic reaction, the rate of heat generation during cement hydration increases with the increase of the ambient temperature. For Class "*G*" Neat System the hydration tests were performed at two different temperatures 77 and 122°F, the density of the slurry was 15.8 ppg (Romero and Loizzo 2000). Figure 22-1 gives the heat production rate per unit of weight as a function of time. We will call the two curves as Ro77 and Ro122. Assessment of the temperature development during cement hydration is necessary to determine how fast the cement will reach an acceptable compressive strength before the casing can be released (Romero and Loizzo 2000). Therefore, for deep wells heat generation during cement hydration has to be taken into account at cement slurry design. The experimental data show that the maximum value of heat generation occurs during the first 5 to 24 hours (Halliburton Cementing Tables 1979).

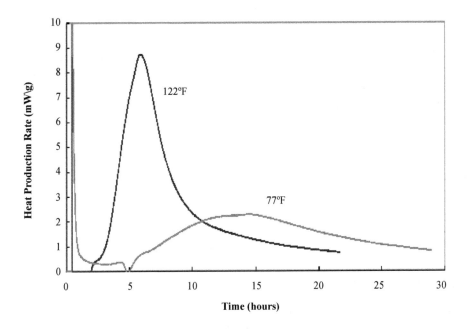

Figure 22-1. Heat production rate per unit of mass as a function of time for class "G" neat cement at two temperatures (Romero and Loizzo 2000).

Thus to determine the amount of heat released by different cement systems during hydration calometric measurements should be completed. The measurements consist of recording the heat flux released by the system while being held at constant temperature. It was found that a quadratic equation (Eq. (22-1)) can be used for a short interval of time to approximate the rate of heat generation (q) per unit of length as a function of time (Kutasov 1999). The objective of this chapter is to demonstrate how field and laboratory data can be utilized to estimate the temperature increase during cement hydration (Kutasov and Eppelbaum 2013):

$$q_D = a_0 + a_1 t + a_2 t^2, \quad q_D = \frac{q^*}{q_r}, \tag{22-1}$$

where a_0, a_1 and a_2 are coefficients, q_r is the reference rate of heat generation per unit of mass, q^* is the rate of heat production per unit of mass, q is the rate of heat production per unit length, and q_D is the dimensionless rate of heat production.

The rate of heat production per unit length is:

$$q = A_0 q_D, \quad A_0 = \pi \left(r_w^2 - r_c^2 \right) \rho_c q_r, \tag{22-2}$$

where A_0 is the reference rate of heat generation per unit of length, r_w is the well radius, r_c is the outside radius of casing, and ρ_c is the density of cement.

Let us assume that from laboratory tests or field tests of cement hydration we are able to determine the peak of the heat production rate—and corresponding time (Figure 22-1). For some small time interval we can assume that a parabola equation approximates the $q_D = q_D(t)$ curve. Then from Eqs. (22-3) and (22-4) we can estimate the coefficients a_1^* and a_2^*:

$$a_1^* t_x + a_2^* t_x^2 = q_{D\max}, \quad \frac{dq_D}{dt} = a_1^* + 2a_2^* t_x = 0, \tag{22-3}$$

$$a_1^* = \frac{2q_{D\max}}{t_x}, a_2^* = -\frac{q_{D\max}}{t_x^2}, \quad q_{D\max} = \frac{q_{\max}}{q_r}, \tag{22-4}$$

where q_{max} is the observed maximum value of heat production rate per unit of mass (at $t = t_x$) and q_{Dmax} is the dimensionless rate of heat production at t_x.

22.2 Working Equation

Below we present a semi-analytical formula which will allow one to estimate the temperature increase versus time (Kutasov 2007). This formula (Eq. (22-4)) describes the transient temperature at the cylinder's wall (T_v), while at the surface of the cylinder the radial heat flow rate (into formations) is a quadratic function of time (Eq. (22-1)):

$$\Delta T = T_V(t) - T_i = \frac{A_0}{2\pi\lambda} \left[a_0 G_0(t) + a_1 G_1(t) + a_2 G_2(t) \right], \tag{22-5}$$

$$G_0(t) = \ln \frac{\left(b + 2d\sqrt{t} \right)^2}{b \left(b + d\sqrt{t} \right)},$$

$$\left\{ \begin{aligned} &G_1(t) = -\ln(b)\left(t + \frac{B}{2}\right) - \frac{1}{2}t \\ &+ \ln\left(b + 2d\sqrt{t}\right)\left(2t - \frac{B}{2}\right) + \ln\left(b + d\sqrt{t}\right)(B - t) \end{aligned} \right\},$$

$$\left\{ \begin{aligned} &G_2(t) = -\ln(b)\left(t^2 - \frac{7}{8}B^2 + Bt\right) - \frac{1}{4}\left(3t^2 + Bt\right) \\ &+ \frac{3}{4}B\sqrt{Bt} + \ln\left(b + 2d\sqrt{t}\right)\left(2t^2 + \frac{1}{8}B^2 - Bt\right) \\ &- \ln\left(b + d\sqrt{t}\right)(t - B)^2 \end{aligned} \right\},$$

$$b = 2.6691, \quad d = \frac{\sqrt{a}}{r_w}, \quad B = \frac{b^2}{d^2},$$

where λ is the thermal conductivity of formation, a thermal diffusivity of formation, and T_i is the static formation temperature.

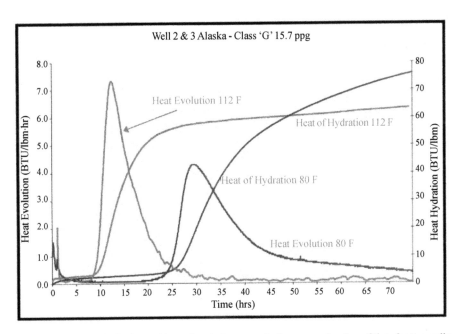

Figure 22-2. Heat of hydration and heat of evolution per unit of mass as a function of time for two wells (Dillenbeck et al. 2002).

From Eq. (22-5) we can estimate the time $t = t_m$ time when the wall temperature reaches its maximum.

$$\frac{dT_v}{dt} = 0, \tag{22-6}$$

or

$$a_0 y_0 + a_1 y_1 + a_2 y_2 = 0, \tag{22-7}$$

$$y_0 = \frac{dG_0}{dt}, \quad y_1 = \frac{dG_1}{dt}, \quad y_2 = \frac{dG_2}{dt}.$$

Software Maple 7 (Waterloo Maple 2001) was utilized to compute derivatives y_0, y_1 and y_2. It was found that

$$y_0 = \frac{\left(\dfrac{2\left(b+2d\sqrt{t}\right)d}{\left(b+d\sqrt{t}\right)b\sqrt{t}} - \dfrac{\left(b+2d\sqrt{t}\right)^2 d}{2\left(b+d\sqrt{t}\right)^2 b\sqrt{t}} \right)\left(b+d\sqrt{t}\right)b}{\left(b+2d\sqrt{t}\right)^2},$$

$$y_1 = \left\{ \begin{array}{l} -\ln(b) - \dfrac{1}{2} + \dfrac{d\left(2t - \dfrac{1}{2}B\right)}{\sqrt{t}\left(b+2d\sqrt{t}\right)} + \\[3mm] 2\ln\left(b+2d\sqrt{t}\right) + \dfrac{d(B-t)}{2\sqrt{t}\left(b+d\sqrt{t}\right)} \\[3mm] -\ln\left(b+d\sqrt{t}\right) \end{array} \right\},$$

$$y_2 = \left\{ \begin{array}{l} -\ln\left(b\right)\left(2t+B\right) - \dfrac{3}{2}t - \dfrac{1}{4}B + \dfrac{3B^2}{8\sqrt{Bt}} \\[3mm] + \dfrac{d\left(2t^2 + \dfrac{1}{8}B^2 - Bt\right)}{\sqrt{t}\left(b+2d\sqrt{t}\right)} + \ln\left(b+2d\sqrt{t}\right)\left(4t-B\right) \\[3mm] -\dfrac{d\left(t-B\right)^2}{2\sqrt{t}\left(b+d\sqrt{t}\right)} - 2\ln\left(b+d\sqrt{t}\right)\left(t-B\right) \end{array} \right\}.$$

22.3 Field Cases

Below we will calculate the transient temperature increase for two cases: (a) rate of heat generation per unit of length at 77°F (curve Ro77, Figure 22-1) and (b) the heat evolution curve at 80°F and it will be referred as A180 (Figure 22-2). The surrounding wellbore formation is oil-bearing sandstone with $\lambda = 1.46$ kcal/(m·hr·°C) and $a = 0.0041$ m²/hr. We selected values of $q_r = 1$ mW/g $= 860.4$ cal/(hr·kg) for the curve Ro77 (Figure 22-1) and $q_r = 1$ BTU/ (lbm·hr) = 553.1 cal/(hr·kg) for the curve A180 (Figure 22-2). In this case the values of heat flow rates per unit of mass will be numerically *equal* to its dimensionless values. To digitize plots and obtain numerical values of q_D and time we used *Grapher* software. The cement retardation time (t_0) was estimated (Table 22-1) from a linear regression program for small values of q_D ($q_D(t = t_0) = 0$ (Figures 22-1 and 22-2). Some input data for our calculations are presented in Table 22-1. Firstly the reference rate of heat generation per unit of length (A_0) was calculated. Secondly the observed values of q_m and t_x were estimated. Then from Eqs. (22-3) and (22-4) the coefficients a_1^* and a_2^* were calculated (Table 22-1). The temperature increase ΔT_x is calculated from Eq. (22-5) (*at* $a_0 = 0$, $a_1 = a_1^*$, $a_2 = a_2^*$). A quadratic regression program was used to process data and the coefficients in Eq. (22-1) were determined. The time (t_m) when the temperature increase reaches its maximum was determined from Eq. (22-7). The corresponding parameters of q_{Dm} and ΔT_{max} were determined from Eqs. (22-1) and (22-5) (Table 22-1).

Table 22-1. Input data and results of calculations.

Input data	Results of calculations
Well Ro77, $4.0 \leq t \leq 14.9$ hours	
$D = 8.5$ in,	$a_1^* = 0.4778$ hr^{-1}, $a_2^* = -0.0249$ hr^{-2}
$D_{oc} = 3.5$ in	$\Delta T_x = 12.8°C$ (Eq. (22-5)) at $a_0 = 0$,
$\rho_c = 15.8$ ppg,	$a_1 = a_1^*, a_2 = a_2^*$
$q_r = 1$ mW/g	$A_0 = 49.73$ Kcal/m·hr, $t_0 = 5$ hrs,
$t_x = 9.59$ hrs	$R = 2.4\%$
$q_{Dmax} = 2.291$	$a_0 = 0.205$, $a_1 = 0.448$ hr^{-1}, $a_2 = -0.0249$ hr^{-2},
$4.0 \leq t \leq 14.9$ hr	$t_m = 12.1$ hrs (Eq. (22-7)), $q_{Dm} = 1.98$ (Eq. (22-1))
$\lambda = 1.46$ kcal·/(m·hr·°C)	at $t = t_m$, $\Delta T_{max} = 13.7°C$
$a = 0.0041$ m²/hr	(Eq. (22-5)) at $t = t_m$
Well A180, $4.0 \leq t \leq 9.8$ hours	
$D = 8.5$ in, $D_{oc} = 3.5$ in	$a_1^* = 1.6696$ hr^{-1}, $a_2^* = -0.1613$ hr^{-2}
$\rho_c = 15.7$ ppg,	$\Delta T_x = 13.2°C$ (Eq. (22-5)) at $a_0 = 0$,
$q_r = 1.$BTU/(lbm·hr)	$a_1 = a^*_1, a_2 = a^*_2$,
$t_x = 5.74$ hrs,	$A_0 = 31.60$ Kcal/m·hr, $t_0 = 24.4$ hrs,
$q_{Dmax} = 4.318$	$R = 1.8\%$
$4.4 \leq t \leq 10.3$ hr	$a_0 = 1.7713$, $a_1 = 0.8810$ hr^{-1}, $a_2 = -0.0757$ hr^{-2},
$\lambda = 1.46$ kcal·/(m·hr·°C)	$t_m = 8.4$ hrs(Eq. (22-7)), $q_{Dm} = 3.83$ (Eq. (22-1))
$a = 0.0041$ m²/hr	at $t = t_m$, $\Delta T_{max} = 15.6°C$
	(Eq. (22-5)) at $t = t_m$

In Table 22-1 the parameters are: $t = t^* - t_0$ is the time since onset of cement hydration, t^* is the time since cement slurry placement, t_0 is the cement retardation time, t_m is the time when the temperature increase reaches its maximum, D is the bit diameter, and D_{oc} is the outside diameter of casing.

The transient values of the temperature increase and dimensionless rate of heat flow rate are presented in Figure 22-3.

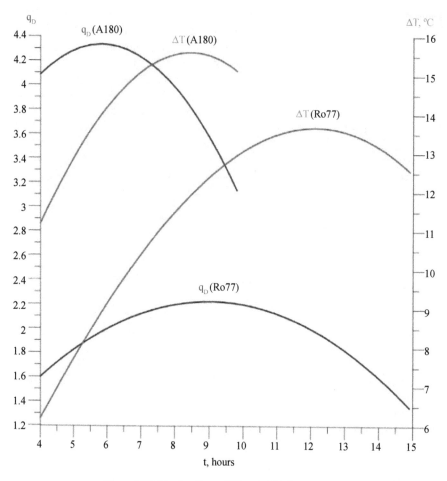

Figure 22-3. The functions $qD(t)$ and $\Delta T(t)$ for two cases.

Thus, the optimal time interval to conduct a temperature survey for well Ro77 is:

$14.6(9.6+5.0) \geq t^* \leq 17.1(12.1+5.0)$ and for well A180,

$30.1(24.4+5.7) \geq t^* \leq 32.8(24.4+8.4)$ hours since cement placement.

　　In conclusion we state that a new method of predicting the temperature increase during cement hydration is suggested. This method can be used to estimate the optimal time to conduct a temperature survey for locating the top of the cement column behind the casing.

23

Cementing of Geothermal Wells– Radius of Thermal Influence

In order to test the mechanical properties of cement under well conditions more closely, the cement must typically be cured or hydrated for the appropriate amount of time under the temperature and pressure conditions that are as close as possible to downhole conditions in the well. At laboratory tests of cement slurries it is important to specify the radius of thermal influence (physical "infinity"). This permits determination of the distance from the axis of the wellbore's model where a constant temperature should be maintained. The above mentioned formula is also used to estimate the radius of thermal influence. An example of calculations is presented.

A semi-analytical formula (Kutasov 2007) was used to estimate the radius of thermal influence. We used this formula and obtained coefficients a_0, a_1 and a_2 to compute the dimensionless radius of thermal influence for the well Ro77 (Figure 23-1). The results of calculations are presented in Figure 23-1.

23.1 Radius of Thermal Influence at Cementing

Earlier we used the thermal balance method to calculate the radius of thermal influence (r_{in}) as a function of time during drilling (Kutasov 1999). Similarly, we will use the thermal balance method to evaluate the value of r_{in} for the cement hydration period (Kutasov and Eppelbaum 2012c). This parameter will allow estimating the degree of thermal disturbance created by the heat of the cement hydration. We found (Kutasov 1999) that the dimensionless temperature distribution around the wellbore during drilling fluid circulation can be approximated by the following equations:

$$\left\{ \begin{array}{l} T_D\left(r_D, t_D\right) = \dfrac{T(r,t) - T_f}{T_w - T_f} = 1 - \dfrac{\ln r_D}{\ln R_{in}}, \\[2mm] R_{in} = \dfrac{r_{in}}{r_w}, \quad r_D = \dfrac{r}{r_w}, \quad 1 \le r_D < R_{in}. \end{array} \right\} \tag{23-1}$$

$$t_D = \frac{at}{r_w^2} = \frac{\lambda t}{c_p \rho r_w^2}. \tag{23-2}$$

and the dimensionless cumulative heat flow rate from the wellbore per unit of length (Q_D) was evaluated:

$$Q_D = \frac{1}{4} \frac{R_{in}^2 - 2\ln(R_{in}) - 1}{\ln(R_{in})}. \tag{23-3}$$

As we mentioned above we obtained a semi-analytical Eq. (22-5) which describes the borehole wall temperature when the borehole is cylindrical source with a quadratic function of the heat flow rate per unit of length.

$$q(t) = A_0\left(a_0 + a_1 t + a_2 t^2\right), \tag{23-4}$$

where A_0 is the reference heat flow rate per unit of length. The cumulative heat flow per unit of length is:

$$Q(t) = \int_0^t q(t)\,dt = A_0 t\left(a_0 + a_1 t/2 + a_2\frac{t^2}{3}\right). \tag{23-5}$$

The cumulative heat flow per unit of length can also be expressed by as:

$$Q = 2\pi\rho c_p r_w^2\left(T_v - T_f\right)Q_D. \tag{23-6}$$

Combining Eqs. (22-5), (23-2), (23-3), (23-5), and (23-6) we obtain an equation to determine the value of R_{in}:

$$\frac{R_{in}^2 - 2\ln(R_{in}) - 1}{4\ln(R_{in})} \cdot \frac{a_0 G_0 + a_1 G_1 + a_2 G_2}{a_0 + \dfrac{a_1}{2}t + \dfrac{a_2}{3}t^2} = t_D. \tag{23-7}$$

23.2 Example of Calculations

The rate of heat generation is presented in Figure 22-1 (Curve Ro77). The surrounding wellbore formation is sandstone. Thermal conductivity is assumed as 1.9 kcal/(m·hr·°C), and thermal diffusivity—as 0.0034 m²/hr. We obtained for $3.0 \le t \le 16.2$ hr:

$$\begin{cases} a_0 = 0.2211, \quad a_1 = 0.4296\,hr^{-1}, \\ a_2 = -0.0228\,hr^{-2}. \end{cases}$$

Please note that the coefficients a_0, a_1, a_2 are slightly different (due to the difference in time intervals) from those in Table 22-1. The input parameters are presented in Table 22-1.

We used the last formula and obtained coefficients a_0, a_1 and a_2 to compute the dimensionless radius of thermal influence for the well Ro77 (Figure 23-1) (Kutasov and Eppelbaum 2011).

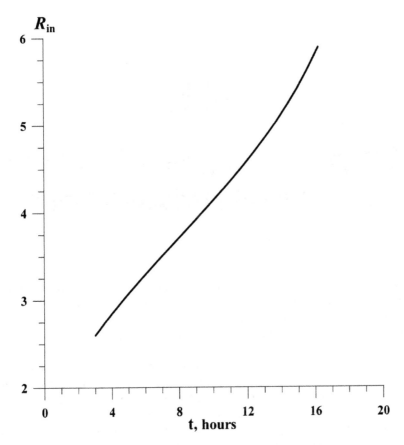

Figure 23-1. The dimensionless radius of thermal influence versus time, wells Ro77, $r_w = 0.108$ m, $a = 0.0041$ m²·hr⁻¹.

Thus in this case at 20 hours of cement hydration at the distance of approximately $6 \cdot 0.108$ m $= 0.65$ m from the well's axis, the ambient temperature should be maintained constant.

24

Temperature Regime of Boreholes: Cementing of Production Liners

The effect of the borehole temperature recovery process affects (disturbed by drilling operations) the technology of the casing cementing operations. The design of cement slurries becomes more critical when a casing liner is used because the performance requirements should be simultaneously satisfied at the top and at the bottom of the liner. For these reasons it is logical to assume that the bottom-hole shut-in temperature should be considered as parameter in the cement slurry design. In deep wells the actual downhole temperature during cement setting may significantly differ from the mud circulating or from the formation (undisturbed) temperatures. In this chapter we suggest two methods (an empirical equation and the equivalent "API Wellbore" method). An early developed relationship can be used to determine the shut-in temperatures. A field example is presented to demonstrate the calculation procedure.

In this chapter we will consider the effect of the borehole temperature recovery process (disturbed by drilling operations) on the production liner cementing casing. Temperature and pressure are two basic influences on the downhole performance (Eppelbaum and Kutasov 2006a) of cement slurries. They affect how long the slurry will pump and how it develops the strength necessary to support the pipe (temperature has the more pronounced influence). The downhole temperature controls the pace of chemical reactions during cement hydration resulting in cement setting and strength development. The shut-in temperature affects how long the slurry will pump and how well it develops the strength to support the pipe. As the formation temperature increases, the cement slurry hydrates and sets faster and develops strength more rapidly. Cement slurries must be designed with sufficient pumping time to provide safe placement in the well. At the same time the cement slurry cannot be overly retarded as this will prevent the development of satisfactory compressive strength. The thickening time of cement is the time that the slurry remains pumpable under set conditions. The specifications of circulation temperature in the design of thickening times for oil well cement are very important. While retarders can extend thickening times, the thickening time for a given concentration of retarder is still very sensitive to changes in temperature. Slurries designed for erroneously high circulating temperatures can have unacceptably long setting times at lower temperatures. A compressive strength of 500 psi (in 24 hours) is usually considered acceptable for casing support (Romero and Loizzo 2000). From Figure 24-1 it follows that a temperature difference of only 6°F (3.3°C) significantly affects the compressive strength development of the cement. To reduce the wait on cement we recommend increasing the outlet mud temperature. Earlier we suggested this technique to reduce wait on cement at surface casing for wells in permafrost

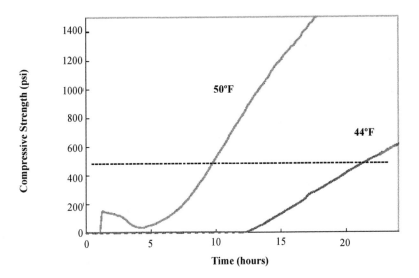

Figure 24-1. Compressive strength development for a deep-water.

regions (Kutasov 1999). This may reduce the cost associated with cementing of the conductor and surface casing.

Formation (undisturbed) temperatures and fluid circulation temperatures at the top of the liner may be much lower than those at the bottom of the liner. Thus, in designing the cement slurry knowledge of the actual temperature to which it is exposed is an important factor. It should be also taken into account that a time lapse exists between the end of mud circulation and placement of the cement.

24.1 Bottom-Hole Circulating Temperatures: Field Data and Empirical Formula

Accurate prediction of bottom-hole circulating temperatures (BHCT) and bottom-hole static temperatures (BHST) is important during the drilling and completion of a well. The downhole temperature is an important factor affecting the thickening time of cement, rheological properties, compressive strength development, and set time.

In 1941, Farris developed bottom-hole temperature charts based on measurements of mud circulating temperatures in five Gulf coast wells. These charts were then used by the American Petroleum Institute (API) to develop a pumping test procedure to predict bottom-hole temperatures during mud circulation (T_{mb}). In 1977 the API task group presented new casing cementing and well-simulation test schedules (New Cement Test Schedules 1977). Forty-one measurements of bottom-hole circulating temperature in water muds were available to estimate the correlation between the values of T_{mb}, geothermal gradient Γ, and vertical depth. Curves were fitted through the measured points for each of the geothermal gradient ranges, thus developing a family of curves. From the curves, schedules were developed to provide laboratory test procedures for simulating cementing of casing (New Cement Test 1977; API Specifications 1982). Although some studies (Venditto and George 1984; Jones 1986) showed an overall agreement with the API schedules, some operators felt that these schedules overestimated the bottom-hole circulating temperatures in deep wells and modified them (Bradford 1985). Later API (Covan and Sabins 1995) has developed new temperature correlations for estimating circulating temperatures for cementing (Table 24-1). To use the current API bottom-hole temperature circulation correlations for designing the thickening time of cement slurries, the average static temperature gradient is required. The surface formation temperature (T_0) for the current API test schedules is assumed to be 80°F.

Table 24-1. The new API temperature correlations (Covan and Sabins 1995).

Depth, ft	Temperature gradient, °F/100 ft					
	0.9	1.1	1.3	1.5	1.7	1.9
8,000	118	129	140	151	162	173
10,000	132	147	161	175	189	204
12,000	148	165	183	201	219	236
14,000	164	185	207	228	250	271
16,000	182	207	233	258	284	309
18,000	201	231	261	291	321	350
20,000	222	256	291	326	360	395

It should also be mentioned that for high geothermal gradients and deep wells, the API circulating temperatures are estimated by extrapolation. Here the current API correlations used to determine the bottom-hole circulating temperature permit prediction in wells with geothermal gradients up to only 1.9°F/100 ft. In an analysis of available field measurements of bottom-hole circulating temperatures (Kutasov and Targhi 1987) it was found that the bottom-hole circulating temperature (T_{mb}) can be approximated with sufficient accuracy as a function of two independent variables: the geothermal gradient, Γ and the bottom-hole static (undisturbed) temperature T_{fb}

$$T_{bot} = d_1 + d_2\Gamma + \left(d_3 - d_4\Gamma\right)T_{fb}. \tag{24-1}$$

For 79 field measurements (Kutasov and Targhi 1987), a multiple regression analysis computer program was used to obtain the coefficients of the formula:

$$d_1 = -50.64°\,C\left(-102.1°\,F\right), \quad d_2 = 804.9\text{m}\ \left(3354\text{ft}\right),$$

$$d_3 = 1.342, \quad d_4 = 12.22\,\text{m}/°\,C\ \left(22.28\ \text{ft}/°\,F\right).$$

These coefficients were obtained for:

$$74.4°C\ \left(166°F\right) \le T_{fb} \le 212.2°C\left(414°F\right),$$

$$1.51°C/100\ \text{m}\ \left(0.83°F/100\text{ft}\right) \le \Gamma \le 4.45\,°C/100\ \text{m}\left(2.44°F/100\text{ft}\right),$$

Therefore, Eq. (24-1) should be used with caution for extrapolated values of T_{fb} and Γ. The accuracy of the results (Eq. (24-1)) is 4.6°C, and was estimated from the sum of squared residuals. The values of T_{fb} can also be expressed as a function of average surface temperature T_0 and total vertical depth (H). For on-land wells the value of T_0 is the formation temperature at a depth of about 20 m.

$$T_{fb} = T_0 + \Gamma\left(H - 20\right). \tag{24-2}$$

For offshore wells the value T_0 is the temperature of bottom sea sediments. It can be assumed that $T_0 \approx 4.4°C$ (40°F) and if the thickness of the water layer is H_w then,

$$T_{fb} = T_0 + \Gamma\left(H - H_w\right). \tag{24-3}$$

In Eqs. (24-2) and (24-3) Γ is the average geothermal gradient. In Table 24-2 the measured and calculated values of bottom-hole circulating temperatures for wells with $T_{fb} > 138°C$ are compared. Measurements 1–7 (West Texas, Gulf Coast of Texas and Louisiana), 8–23 (Texas, Louisiana), 24, 25–27 (Mississippi) and 28–32 (Table 24-2) were taken from Shell and Tragesser (1972); Venditto and George (1984); Jones (1986); Wooley et al. (1984) and Sump and Williams (1973) respectively.

24.1.1 Comparison with API schedules

In Table 24-3 the current API schedules (Covan and Sabins 1995) and values of T_{mb} calculated by Eq. (24-1) are compared. It can be concluded that for deep wells and high temperature gradients the API bottom-hole circulating temperatures are too high. We recommend the use of Eq. (24-1) for estimating the bottom-hole circulating temperature for high geothermal gradients. It is possible that the coefficients in Eq. (24-1) should be corrected to account for very high geothermal gradients which are common in geothermal wells.

Table 24-2. Measured T^*_{mb} and predicted T_{mb} bottom-hole circulating temperatures. T_{fb} is the bottom-hole static temperature, $\Delta T = T^*_{mb} - T_{mb}$.

No	Well	H, m	Γ, °C/100 m	T_{fb}, °C	T^*_{mb}, °C	T_{mb}, °C	ΔT, °C
1	8	6050	3.01	205	177	173	4
2	9	6926	2.08	167	146	147	−1
3	10	4237	4.45	212	145	154	−9
4	12	4876	3.14	176	146	143	3
5	16	3747	3.70	162	121	123	−2
6	16	3461	3.44	143	112	109	3
7	16	4887	3.30	185	140	149	−9
8	4	3962	3.37	160	130	125	5
9	5	4454	3.24	171	134	137	−3
10	16	5333	2.46	158	128	133	−5
11	22	4206	2.90	148	118	119	−1
12	27	3352	3.54	145	106	109	−3
13	27	3535	3.52	151	113	115	−2
14	35	3627	3.68	160	131	121	10
15	36	5427	2.48	161	131	136	−5
16	38	5056	2.64	160	138	133	5
17	38	5529	2.57	168	148	143	5
18	38	5898	2.68	184	165	158	7
19	40	3266	3.44	139	110	105	5
20	46	4389	3.17	165	141	132	9
21	47	4079	3.63	175	137	135	2
22	47	4518	3.43	181	142	144	−2
23	51	3806	3.30	152	121	118	3
24	4	3718	3.28	145	112	112	0
25	MS	4900	2.72	153	129	126	3
26	MS	6534	2.55	187	163	162	1
27	MS	7214	2.57	206	178	182	−4
28	1	4578	2.86	157	137	128	9
29	2	4971	2.95	169	142	139	3
30	3	4571	2.92	156	123	127	−4
31	8	5486	3.46	211	179	171	8
32	9	6926	2.11	167	146	148	−2

Table 24-3. Bottom-hole circulating temperatures. The values on the first line are predicted values by empirical Eq. (24-1); second line is from current API schedules (Covan and Sabins 1995).

Depth, ft	Geothermal gradient, °F/100 ft								
	0.9	1.1	1.3	1.5	1.7	1.9	2.1	2.3	2.5
4,000	-	-	-	-	-	100	107	113	119
	99	100	101	103	104	-	-	-	-
6,000	-	-	102	114	125	135	143	151	158
	99	100	101	102	103	104	-	-	-
8,000	-	113	129	144	158	170	180	189	197
	118	129	140	151	162	173	-	-	-
10,000	116	137	157	174	190	205	217	227	236
	132	147	161	175	189	204	-	-	-
12,000	136	161	184	205	223	239	254	226	276
	148	165	183	201	219	236	-	-	-
14,000	157	185	211	235	256	274	290	304	315
	164	185	207	228	250	270	-	-	-
16,000	178	210	239	265	289	309	327	342	354
	182	207	233	258	284	309	-	-	-
18,000	198	234	266	295	321	344	364	380	393
	201	231	261	291	321	350	-	-	-
20,000	219	258	293	326	354	379	400	418	433
	222	256	291	321	360	395	-	-	-

24.1.2 The equivalent "API Wellbore" method

As was mentioned earlier, the American Petroleum Institute (API), Sub-committee 10 (Well Cements) has developed new temperature correlations for estimating circulating temperatures for cementing (Covan and Sabins 1995; API RP 10B, Recommended Practice 1997). To use the current API bottom-hole temperature circulation (BHCT) correlations (schedules) for designing the thickening time of cement slurries (for a given depth) the average static temperature gradient needs to be known. The surface formation temperature (SFT) for the current API test schedules is assumed to be 80°F. Thus to calculate the static temperature gradient, the static (undisturbed) temperature profile of formations should be determined with reasonable accuracy. A value of SFT (the undisturbed formation temperature at a depth of approximately 50 ft, where the temperature is practically constant) of about 80°F is typical only for wells in the southern U.S. and some other regions. For this reason the API test schedules cannot be used to calculate values of BHCT for cementing in wells drilled in deep waters, in areas remote from the tropics, or in Arctic regions. For example, the equivalent parameter of SFT for offshore wells is the temperature of sea bottom sediments (mud line) close to 40°F. In Arctic areas the value of SFT is well below the freezing point of water. Many drilling operators came to the conclusion that computer temperature simulation models (instead of the API schedules) should be used to estimate the cementing temperatures (Honore et al. 1993; Guillot et al. 1993; Calvert and Griffin 1998). Below we present a novel concept—the Equivalent "API Wellbore Method" (Kutasov 2002) and show that the current API bottom-hole temperature circulation (BHCT) correlations can be used for any deep well and for any values of surface formation temperature. We call this technique the "API-EW Method". An empirical formula and the results of computer simulations were utilized to verify the applicability of the technique.

It was shown that the bottom-hole circulating temperature can be approximated with sufficient accuracy by an empirical Kutasov-Targhi (KT-Formula) (Eq. (24-1)) as a function of two independent variables: the static temperature gradient (Γ) and the bottom-hole static (undisturbed) temperature (T_{fb}). The values of T_{fb} can also be expressed as a function of surface formation temperature T_0 and total vertical depth (H). As was mentioned above, for on-land wells the value of T_0 is the temperature of formations at a depth of about 50 ft.

$$T_{f0} = T_0 + \Gamma(H - 50).$$

In practice, for deep wells it is usually assumed that

$$T_{f0} = T_0 + \Gamma H. \tag{24-4}$$

For offshore wells the value T_0 is the temperature of bottom sea sediments. It can be assumed that $T_0 \approx 40°F$ and if the thickness of the water layer is H_w, then

$$T_{f0} = T_0 + \Gamma(H - H_w). \tag{24-5}$$

24.1.3 The "API-EW method"

Note that the API bottom-hole circulation temperature correlations are based on field measurements in many deep wells. To process field data, the staff of the API Sub-Committee 10 set two variables—the average static temperature gradient and the vertical depth. The problem is in assuming a constant value of the surface formation temperature. In fact, to use the API schedules, the drilling engineer has to estimate the static temperature gradient from the following formula:

$$\Gamma = \frac{T_{fb} - 80}{H}. \tag{24-6}$$

The reader can see the difference between relationships (24-4) and (24-5) and the last formula. This is why significant deviations between measured and predicted values of BHCT were observed (Table 24-3). It is logical to assume that for wells with $T_0 = 80°F$ good agreement between measured and estimates from API correlations values of BHCT should be expected. Therefore we suggest "transforming" a real wellbore into an "Equivalent API Wellbore". For purposes of illustration, let us consider a well with following parameters: $H = 20,000$ ft, $\Gamma = 0.020°F/ft$ and $T_0 = 60°F$. The depth of the 80°F isotherm is: $(80–60)/0.020 = 1,000$ (ft). Thus the vertical depth of the "Equivalent API Wellbore" is $H^* = 20,000–1,000 = 19,000$ (ft).

Table 24-4. Range of errors obtained with the simulation and the new API method (Guillot et al. 1993).

	Standard deviation, °F	Maximum overestimation, °F	Maximum underestimation, °F
New API	18.3	36	−29
Simulator	10.9	17	−19

Similarly, for a well with $T_0 = 100°F$,

$$H^* = 20,000 + 1,000 = 21,000 \, (\text{ft}).$$

Below we present simple equations to estimate the equivalent vertical depth (H^*). For an on-land well,

$$T_{fb} = T_0 + \Gamma H = 80 + \Gamma H^*. \tag{24-7}$$

$$H^* = H + \frac{T_0 - 80}{\Gamma}. \tag{24-8}$$

For an offshore well

$$H^* = (H - H_w) + \frac{T_0 - 80}{\Gamma}, \tag{24-9}$$

where T_0 is the temperature of bottom sediments (mud line) and Γ is the average temperature gradient in the $H - H_w$ section of the wellbore.

$$\Gamma = \frac{T_{fb} - T_0}{H - H_w}. \tag{24-10}$$

Examples

Below we present six examples of ways to calculate the bottom-hole circulating temperatures (BHCT) by the API-EW Method.

Example 1: on land well A (Figure 24-2). The surface temperature $T_0 = 50°F$, vertical depth $H = 20,000$ ft, and the static temperature gradient $\Gamma = 0.015°F/ft$, and the bottom-hole static temperature is 350°F.

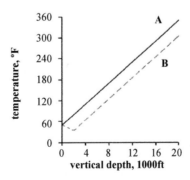

Figure 24-2. Temperature profiles: (A) on land well, (B) offshore well.

From Eq. (24-8) we obtain

$$H^* = 20,000 + \frac{50-80}{0.015} = 18,000\,(\text{ft}).$$

Using the new API BHCT correlations (Covan and Sabins 1995) for $H^* = 18,000$ ft and $\Gamma = 0.015°F/ft$ we estimate that BHCT $= 291°F$ and this is in fairly good agreement with the value of $301°F$ determined from Eq. (24-1) (note that the average accuracy of Eq. (24-1) is $8.2°F$).

Example 2: offshore well B (Figure 24-2). The water surface temperature is $50°F$, the temperature of bottom sediments $T_0 = 35°F$, $H = 20,000$ ft, $\Gamma = 0.015°F/ft$, water depth $H_w = 2,000$ ft, and the bottom-hole static temperature is $305°F$.

From Eq. (24-10) we obtain

$$H^* = (20,000 - 2,000) + \frac{35-80}{0.015} = 15,000\,(\text{ft}).$$

Based on the new API correlations (Covan and Sabins 1995) for $H^* = 15,000$ ft and $\Gamma = 0.015°F/ft$ we found that the BHCT $= 244°F$ which is consistent with the value of $256°F$ calculated from Eq. (24-1).

Example 3: offshore well (Calvert and Griffin 1998). The sea-surface temperature is $76°F$, the temperature of bottom sediments (mud-line temperature: $T_0 = 38°F$, $H = 10,125$ ft, water depth $H_w = 3,828$ ft, and the bottom-hole static temperature is $180°F$. In this case

$$\Gamma = \frac{180-38}{10125-3828} = 0.02255\,(°F/ft).$$

From Eq. (24-10) we obtain

$$H^* = (10,125 - 3,828) + \frac{38-80}{0.02255} = 4,434\,(\text{ft}).$$

Using the new API correlations (Covan and Sabins 1995) for $H^* = 4,434$ ft and $\Gamma = 0.02255°F/ft$ we determined (extrapolation was used) that BHCT $= 116°F$ which is in line with the results of computer simulations (Table 24-5). The depth of $H^* = 4,434$ ft is beyond the range of applicability of Eq. (24-1) and for this reason we did not use this formula.

Table 24-5. Results of simulations and calculations of bottom-hole circulating temperatures (Calvert and Griffin 1998, modified).

Model	BHCT, °F
Simulator A	99
Simulator B	108
API (Spec 10, 1990)	144
API (RP 10B, 1997)	140
API–EW	116

Examples 4, 5, 6: the parameters for three wells (cases) were taken from an article by Goodman et al. (1988). The results of calculations and computer simulations are presented in Table 24-6. It shows that the API-EW Method predicts the bottom-hole circulating temperatures with satisfactory accuracy. The average deviation from the computer stimulation results (for three cases) was 11°F.

Table 24-6. Results of simulations and calculations of bottom-hole circulating temperatures.

Parameters	Well 2	Well 6	Well 8
TVD, ft	15,000	15,000	11,000
Water Depth, ft	0	1,000	1,000
Equivalent TVD, ft	15,000	12,000	8,000
Surface Temp., °F	80	80	80
Seabed Temp., °F	-	50	50
Static Gradient, °F/ft	0.015	0.015	0.015
BHST, °F	305	260	200
BHCT: API-EW, °F	244	201	140
BHCT: Simulator, °F	248	189	157
BHCT: KT-Formula, °F	255	210	150

24.2 The Shut-in Temperature

By using the adjusted drilling mud circulation time concept (see Chapter 2) we obtained an equation for dimensionless shut-in temperature (Kutasov 1999):

$$T_{sD} = \frac{T_s(t_s) - T_f}{T_c - T_f}, \tag{24-11}$$

$$\left\{ T_{sD} = 1 - \frac{Ei\left[-\beta\left(1 + t_D^* / t_{sD}\right)\right]}{Ei(-\beta)}, \quad t_{sD} = \frac{at_s}{r_w^2} \right\}, \tag{24-12}$$

$$t_D = \frac{at_c}{r_w^2}, \quad \beta = \frac{1}{4t_D^*}, \quad t_D^* = Gt_D, \tag{24-13}$$

where T_s is the shut-in temperature, t_s is the shut-in time, T_c is the drilling fluid circulation temperature (at a given depth), t_c is the drilling fluid circulation time, and T_f is the formation (undisturbed) temperature.

The values of the function G can be calculated from Eqs. (2-1) and (2-2) (see Chapter 2).

24.2.1 Field example

Well #4 (Venezuela) is a vertical wellbore. The total depth was 12,900 ft the bottom-hole static temperature at 12,600 ft was 244°F. The casing size of this well is $5^{1/2}$ in.

And the hole size was 8 1/2-in. = 0.354 ft the 14.0 ppg composite blend cement slurry was used (Dillenbeck et al. 2002). We assumed that the surrounding formation is oil-bearing sandstone with thermal conductivity of 1.46 kcal/(m·hr·°C) and thermal diffusivity of 0.0041 m²/hr. Let us assume that a production liner was set at a depth of 10,000 to 12,600 ft. A period of 30 days was needed to drill this interval. In addition to this, 36 hours were needed to clean up the well and pump the cement slurry; therefore, the duration of the thermal disturbance of the formation at depth of 10,000 ft was 756 hrs [(30·24) + 36] days and at depth of 12,600 ft was 36 hours. Let us assume that the geothermal gradient Γ = 0.013°F/ft. Then the formation temperature is

$$T_f\left(°F\right)= 80\left(°F\right)+0.013\left(°F/ft\right)\cdot H.$$

The drilling mud circulating temperatures were estimated from Eq. (24.1): for H = 12,600 ft, T_{mb} = 199°F, and for H = 10,000 ft, T_{mb} = 162.5°F.

The results of calculations after Eq. (24-12) are presented in Figure 24-3 and Table 24-7.

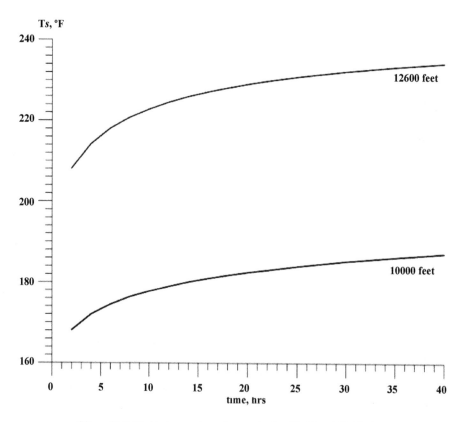

Figure 24-3. Shut-in temperatures for two depths: 10,000 and 12,600 feet.

The temperature variation (Table 24-7) and Figure (24-3) depends on the time of the cement slurry placement and can have a significant effect on the performance of a cement system and has to be taken into account. For example, the temperature at depth of 10,000 ft after 6 hours is not 162.5°F (circulation temperature), but 174.5°F. Also, the temperature at the depth 12,600 ft after 6 hours of shut-in is not 199°F (circulation temperature), but 218.1°F. The difference of 12–19°F can have a substantial impact on the cement slurry design. For both depths the difference of $T_f - T_s$ after 6 hours of shut-in is 25.9 and 35.5°F (Table 24-7).

Table 24-7. The difference $T_f - T_s$ for two depths.

t_s,hrs	$h = 12,600$ ft $t_c = 36$ hrs $T_f = 244°F$ $T_c = 199°F$	$h = 10,000$ ft $t_c = 756$ hrs $T_f = 210°F$ $T_c = 162.5°F$	
	$T_f - T_s$, °F	$T_f - T_s$,°F	
2.0	35.9	41.9	
4.0	29.8	38.0	
6.0	25.9	35.5	
8.0	23.2	33.6	
10.0	21.1	32.2	
14.0	18.0	29.9	
18.0	15.8	28.2	
20.0	14.9	27.5	
30.0	11.7	24.7	
40.0	9.7	22.8	

25

Recovery of the Thermal Equilibrium in Deep and Super Deep Wells: Utilization of Measurements While Drilling Data

The modelling of primary oil production and design of enhanced oil recovery operations, well log interpretation, well drilling and completion operations, and evaluation of geothermal energy resources require knowledge of the undisturbed reservoir temperature. Temperature measurements in wells are mainly used to determine the temperature of the Earth's interior. The drilling process, however, greatly alters the temperature of the reservoir immediately surrounding the well. The temperature change is affected by the duration of drilling fluid circulation, the temperature difference between the reservoir and the drilling fluid, the well radius, the thermal diffusivity of the formations, and the drilling technology used. Given these factors, the exact determination of reservoir temperature at any depth requires a certain length of time in which the well is not in operation. In theory, this shut-in time is infinitely long. There is, however, a practical limit to the time required for the difference in temperature between the well wall and surrounding reservoir to become a specified small value (Kutasov and Eppelbaum 2011).

The results of field and analytical investigations have shown that in many cases the effective temperature (T_w) of the circulating fluid (mud) at a given depth can be assumed constant during drilling or production (Lachenbruch and Brewer 1959; Ramey 1962; Edwardson et al. 1962; Jaeger 1961; Kutasov et al. 1966; Raymond 1969; Kutasov 1999; Eppelbaum et al. 2014). Here we should note that even for a continuous mud circulation process the wellbore temperature is dependent on the current well depth and other factors. The term "effective fluid temperature" is used to describe the temperature disturbance of formations while drilling. In their classical work Lachenbruch and Brewer (1959) have shown that the wellbore shut-in temperature mainly depends on the amount of thermal energy transferred to (or from) formations.

While drilling deep sections of super deep wells the penetration rates become small and the time of thermal disturbance increases. It is demonstrated that minimum field data: records of stabilized outlet mud temperature and several values of bottom-hole mud temperature measured while drilling (MWD) are needed to construct an empirical equation which approximates the downhole temperature profile during drilling. An analytical equation is presented which describes recovery of the thermal equilibrium when the temperature of drilling fluid (at a given depth) is a linear function of time.

Below it will be shown that in deep wells the mud circulation temperature can be approximated as a linear time function. The objective of this chapter is twofold: (a) to demonstrate how temperature measurements while drilling (MWD) can be utilized to determine the downhole drilling mud temperature profile, and, (b) how the MWD data can be used in estimation of the transient shut-in temperatures. Calculations of shut-in temperatures for two field examples are also presented.

25.1 Empirical Equation

The temperature surveys in many deep wells have shown that both the outlet drilling fluid temperature and the bottom-hole temperature varies monotonically with the vertical depth (Figure 25-1). It was suggested (Kuliev et al. 1968) that the stabilized circulating fluid temperature in the annulus (T_m) at any point can be expressed as

$$T_m = A_0 + A_1 h + A_2 H, \quad h < H, \tag{25-1}$$

where the values A_0, A_1, and A_2 are constants for a given area, h is the current vertical depth and H is the total vertical depth of the well (the position of the bottom of the drill pipe at fluid circulation). The values of A_0, A_1, and A_2 are dependent on drilling technology (flow rate, well design, fluid properties, penetration rate, etc.), geothermal gradient and thermal properties of the formation. It is assumed that, for the given area, the above mentioned parameters vary within narrow limits. In order to obtain the values of A_0, A_1, and A_2 the records of the outlet fluid (mud) temperature (at $h = 0$) and results of downhole temperature surveys are needed. In Eq. (25-1) the value of T_m is the stabilized downhole circulating temperature. The time of the downhole temperature stabilization (t_s) can be estimated from the routinely recorded outlet mud temperature logs (Kutasov et al. 1988; Kutasov 1999).

Eq. (25-1) was verified (Kutasov et al. 1988) with more than 10 deep wells, including two offshore wells, and the results were satisfactory. Here we are presenting two examples of applying Eq. (25-1) for predicting downhole circulating temperatures. It will be shown that only a minimum of field data is needed to use this empirical method.

25.1.1 Mississippi well

The results of field temperature surveys and additional data (Table 25-1) were taken from the paper by Wooley et al. (1984). Three measurements of stabilized bottom-hole circulating temperatures and three values of stabilized outlet mud temperatures

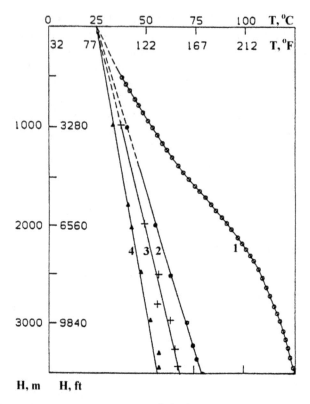

Figure 25-1. Temperature curves versus current well depth. Well 12-PXC, Stavropol district, Russia (Proselkov 1975). (1) geothermal curve, (2) circulating bottom-hole temperature, (3) outlet drilling mud temperature, (4) inlet drilling mud temperature.

were run in a multiple regression analysis computer program and the coefficients of the empirical Eq. (25-1) were obtained

$$A_0 = 32.68 °C, A_1 = 0.01685° C/m, A_2 = 0.003148° C/m.$$

Thus, an equation for the downhole circulating temperature is

$$T_m = 32.68 + 0.01685h + 0.003148H. \tag{25-2}$$

25.1.2 Webb County, Texas

The temperature measurements (Table 25-1) in this location were obtained from the paper by Venditto and George (1984). It was not known whether these measurements were taken in a single well or in the wells in the same area. But since this empirical method can be applied to an entire area as well as to a single well, the data points were used simultaneously to calculate the coefficients in Eq. (25-1). By using the multiple regression analysis a computer program was obtained,

$$A_0 = 5.69°C, A_1 = 0.00636°C/m, A_2 = 0.01714°C/m.$$

Table 25-1. Measured ($T_m{}^*$) and predicted (T_m) values of wellbore circulating temperature (Kutasov 1999).

h, m	H, m	$T_m{}^*$, °C	T_m, °C	$T_m{}^* - T_m$, °C
Mississippi well				
4900	4900	129.4	130.7	−1.3
6534	6534	162.8	163.4	−0.6
7214	7214	178.3	177.0	1.3.6
0	4900	50.0	48.1	1.9
0	6534	51.7	53.2	−1.5
0	7214	55.6	55.4	0.2
Webb County, Texas				
2805	2805	70.6	71.6	−1.0
3048	3048	78.3	77.3	1.0
3449	3449	86.7	86.7	0.0
0	2805	53.3	53.8	−0.5
0	3048	60.0	57.9	2.1
0	3261	60.0	61.6	−1.6

Thus, the equation for the downhole circulating temperature is

$$T_m = 5.69 + 0.0036h + 0.01714H. \tag{25-3}$$

In Table 25-1 the measured and predicted values of bottom-hole and outlet circulating temperatures are compared and the agreement is seen to be good in both cases. The significant difference in values of A_0, A_1, and A_2 for the Mississippi and the Texas wells indicates that these coefficients are valid only within a given area.

Let us assume that for the well section $(H - h)$ the penetration rate is constant (u). Then $H = h + ut$ and from Eq. (25-1):

$$T_m = A_0 + A_1 h + A_2 (h + ut)$$

or

$$T_m = B_0 + B_1 t, \qquad B_0 = A_0 + h(A_1 + A_2), \qquad B_1 = A_2 u. \tag{25-4}$$

Introducing the dimensionless circulation time (t_D) we obtain

$$T_m = B_0 (1 + b t_D), \qquad t_D = \frac{a t_c}{r_w^2}, \qquad b = \frac{B_1 r_w^2}{B_0 a}. \tag{25-5}$$

25.2 Cumulative Heat Flow

Constant drilling mud temperature. As we mentioned earlier, the wellbore shut-in temperature mainly depends on the amount of thermal energy transferred to (or from) formations (Lachenbruch and Brewer 1959). It is known that the cumulative heat flow from the wellbore per unit of length is given by:

$$Q = 2\pi\rho\, c_p r_w^2 \left(T_w - T_f\right) Q_D\left(t_D\right),\tag{25-6}$$

where $T_w = T_m$ is the temperature of the drilling fluid (at a given depth), ρ is the density of formations, c is the specific heat of formations, r_w is the well radius, and Q_D is the dimensionless cumulative heat flow. The time dependent function Q_D can be obtained from the integral

$$Q_D = \int_0^{t_D} q_D\left(t_D\right) dt_D,\tag{25-7}$$

where q_D is the dimensionless heat flow rate. We found (Kutasov 1987) that the dimensionless heat flow rate can be approximated by a semi-analytical equation:

$$q_D = \frac{1}{\ln\left(1 + D\sqrt{t_D}\right)},\tag{25-8}$$

$$D = d + \frac{1}{\sqrt{t_D} + b}, \quad d = \frac{\pi}{2}, \quad b = \frac{2}{2\sqrt{\pi} - \pi}, \quad b = 4.9589.\tag{25-9}$$

The dimensionless flow rate (q_D) can be also determined by using empirical Eq. (25-10) (Chiu and Thakur 1991)

$$q_D = \frac{1}{c_1 \ln\left[1 + c_2\sqrt{t_D}\right]}, \quad c_1 = 0.982, \quad c_2 = 1.81.\tag{25-10}$$

Software Maple 7 (Waterloo Maple 2001) was utilized to compute the integral Q_D, where the function q_D is given by Eq. (25-10). It was found that

$$Q_D\left(t_D\right) = -2\frac{Ei\left(1, -2u\right) - Ei\left(1, -1u\right) - 0.69315}{c_1 c_2^2}, \quad u = \ln\left(1 + c_2\sqrt{t_D}\right),\tag{25-11}$$

where $-Ei(1, -x) = Ei(+x)$ is the exponential integral of a positive argument. In Table 25-2 the values of Q_D computed by two numerical integration methods are compared. The agreement between values of Q_D calculated by these two methods is seen to be good.

25.3 Drilling Mud Temperature as Linear Time Function

When the drilling mud temperature can be approximated by a linear function (Eq. (25-4)) the Duhamel integral can be used

$$Q_{Dt} = \int_0^{t_D} T_{wt}\left(\lambda\right) \cdot \frac{d}{dt_D} Q_D\left(t_D - \lambda\right) d\lambda,\tag{25-12}$$

Table 25-2. Comparison of values of dimensionless cumulative heat flow rate for a well with constant bore-face temperature. Q_D^* is determined by (Van Everdingen and Hurst 1949); Q_D is computed by the use of Eq. (25-11).

t_D	Q_D^*	Q
2	0.2447E + 01	0.2448E + 01
3	0.3202E + 01	0.3204E + 01
5	0.4539E + 01	0.4540E + 01
10	0.7411E + 01	0.7411E + 01
20	0.1232E + 02	0.1233E + 02
50	0.2486E + 02	0.2485E+ 02
100	0.4313E + 02	0.4302E + 02
200	0.7579E + 02	0.7555E + 02
500	0.1627E + 03	0.1606E + 03
1000	0.2935E + 03	0.2922E + 03
2000	0.5341E + 03	0.5317E + 03
5000	0.1192E + 04	0.1187E + 04
10000	0.2204E + 04	0.2195E + 04
20000	0.4096E + 04	0.4083E + 04
50000	0.9363E + 04	0.9342E + 04
100000	0.1759E + 05	0.1756E + 05

where

$$T_{wt} = \frac{T_m}{T_o} = (1 + bt_D), \quad T_{wt}(\lambda) = 1 + b\lambda$$

and Q_D is the dimensionless cumulative heat flow rate for a well with constant bore-face temperature and λ is the variable of integration. Utilization of Eq. (25-11) does not allow the integration of the Duhamel Integral. For this reason we used a simple function (Eq. (25-13)):

$$Q_D = At_D^c \tag{25-13}$$

to approximate the results of a numerical solution (Van Everdingen and Hurst 1949). A linear regression program was used to compute the coefficients A and C (Table 25-3).

For $t_D > 100$ with $R = \dfrac{\Delta Q}{Q} < 1.5\%$ the values of coefficients C and A can be approximated by the following expressions:

$$c = 0.028704 \cdot \ln t_D + 0.65297, a = -0.148 \cdot \ln t_D + 1.837. \tag{25-14}$$

Table 25-3. Coefficients C and A.

t_D	Number of points	C	A	R, %
5–100	22	0.75361	1.3088	0.68
100–400	15	0.82297	0.97022	0.05
400–1000	25	0.85013	0.82569	0.01
1000–1500	21	0.86236	0.75935	0.00
1500–2000	11	0.86735	0.73392	0.04
2000–3000	21	0.87274	0.70274	0.01
3000–5000	41	0.88213	0.646075	0.01

Now the integral (Eq. (25-12)) can be evaluated

$$Q_{Dt} = At_D^c \left[1 + bt_D - \frac{bt_D}{(c-1)(c+1)} \right], \qquad c < 1. \tag{25-15}$$

25.4 Radius of Thermal Influence

In theory the drilling process affects the temperature field of formations at very long radial distances. There is, however, a practical limit to the distance—the radius of thermal influence (r_{in}), where for a given circulation period $(t = t_c)$ the temperature $T(r_{in}, tc)$ is practically equal to the geothermal temperature T_f. To avoid uncertainty, however, it is essential that the parameter r_{in} must not be dependent on the temperature difference $T(r_{in}, t_c) - T_f$. For this reason we used the thermal balance method to calculate the radius of thermal influence.

The results of modeling, experimental works, and field observations have shown the temperature distribution around the wellbore during drilling can be approximated by the following relation (Kutasov 1999):

$$\frac{T(r,t) - T_f}{T_w - T_f} = 1 - \frac{\ln \dfrac{r}{r_w}}{\ln \dfrac{r_{in}}{r_w}}, \qquad r_w \le r \le r_{in}. \tag{25-16}$$

Introducing the dimensionless values of circulation time, radial distance, radius of thermal influence, and temperature

$$t_D = \frac{at_c}{r_w^2}, \quad r_D = \frac{r}{r_w}, \quad R_{in} = \frac{r_{in}}{r_w}, \quad T_D(r_D, t_D) = \frac{T(r,t) - T_f}{T_w - T_f} \tag{25-17}$$

we obtain

$$T_D(r_D, t_D) = 1 - \frac{\ln r_D}{\ln R_{in}}, \qquad 1 \le r_D \le R_{in}, \tag{25-18}$$

The dimensionless cumulative heat flow per unit of length is given by:

$$Q_D = \int_0^{t_D} q_D \, dt_D = \int_1^{R_{in}} T_D \left(r_D, t_D \right) r_D \, dr_D. \tag{25-19}$$

The last integral is evaluated by using the table for the following integral (Gradshtein and Ryzhik 1965)

$$\int x^n \ln x \, dx = \frac{x^{n+1}}{n+1} \left(\ln x - \frac{1}{n+1} \right). \tag{25-20}$$

From Eqs. (25-18)–(25-20) we obtain

$$Q_D = \frac{R_{in}^2 - 1}{4 \ln R_{in}} - \frac{1}{2}. \tag{25-21}$$

By equating values of Q_{Di} and Q_D (Eqs. (25-15) and (25-21)) we obtain a function $R_{in} = f(t_D)$ which can be used to determine the dimensionless radius of thermal influence

$$At_D^c \left[1 + bt_D - \frac{bt_D}{(c-1)(c+1)} \right] = \frac{1}{4} \frac{R_{in}^2 - 2 \ln \left(R_{in} \right) - 1}{\ln \left(R_{in} \right)}. \tag{25-22}$$

25.5 Shut-in Temperature

Thus, for the moment of time $t = t_c$ the temperature distribution in and around the wellbore is

$$\begin{aligned}
T_{wt} &= T_m = B_0 + B_1 t_c, & 0 &\le r_D \le 1, \\
T(t_c, r), \text{ Eq. (25-16)}, & & 1 &\le r_D \le R_{in}, \\
T_f, & & r_D &> R_{in}.
\end{aligned} \tag{25-23}$$

For the temperature distribution (Eq. (25-23)) we obtained the following formula for the wellbore shut-in temperature T_s (Kutasov 1999):

$$\frac{T_s \left(t_s, 0 \right) - B_0 - B_1 t_c}{T_f - B_0 - B_1 t_c} = \frac{Ei \left[-p(R_{in})^2 \right] - Ei \left(-p \right)}{2 \ln R_{in}}, \tag{25-24}$$

$$p = \frac{1}{4nt_D}, \quad n = \frac{t_s}{t_c},$$

$$T_f = \frac{2 \left(T_s - B_0 - B_1 t_c \right) \ln R_{in}}{Ei \left[-p(R_{in})^2 \right] - Ei \left(-p \right)} + B_0 + B_1 t_c. \tag{25-25}$$

Hence, when the values of B_0 and B_1 are known, only one value of shut-in temperature (T_s) is needed to determine the undisturbed formation temperature (T_f).

The derivation of Eq. (25-24) assumes that the difference in thermal properties of drilling mud and formations can be neglected. Although this is a conventional assumption even for interpreting bottom-hole temperature surveys (Timko and Fertl

1972; Dowdle and Cobb 1975), when the circulation periods are small, Eq. (25-24) should be used with caution for very small shut-in times.

25.6 Field Examples

25.6.1 Mississippi well (Wooley et al. 1984)

50 days were spent to drill the 6,534-7, 214 m section of this well. Thus the average penetration was $u = 0.566$ m/hr and the values of B_1 and B_0 were estimated (see Eq. (25-4)):

$$B_1 = 0.001783 \,°C/hr, \qquad B_0 = 163.6 \,°C.$$

The undisturbed temperature of formations at $h = 6,534$ m is $T_f = 187.8°C$, the well radius $r_w = 0.0984$ m, and the coefficient of thermal diffusivity $a = 0.0040$ m²/hr (assumed). The average temperature of the drilling mud at this depth is $T_f = 163.6°C$ (Eq. (25-2)). The radius of thermal influence was computed from Eq. (25-22).

25.6.2 Webb County, Texas. Well #30 (Venditto and George 1984)

The total vertical depth is 10,000 ft. For comparison purposes we will assume that while drilling the 8,000–10,000 ft section of the well the average penetration was $u = 0.566$ m/hr, and $r_w = 0.0984$ m, $a = 0.0040$ m²/hr. In this case $t_c = 1,077$ hrs. The values of B_1 and B_0 were estimated (see Eq. (25-4)):

$$B_1 = 0.009701°C/hr, \qquad B_0 = 63.0 \,°C.$$

The undisturbed temperature of formations at $h = 8,000$ ft m is $T_f = 91.6°C$. The average temperature of the drilling mud at this depth is $T_f = 68.2°C$ (Eq. 25-3). The radius of thermal influence was computed by the use of Eq. (25-22).

In Table 25-4 we present results calculations after Eq. (25-24) values of $\Delta T = T_f - T_s$. We also consider the case when the average temperature of the drilling mud (during the circulation period) is used at calculations of ΔT. For the Well #30 we present results of calculations when the penetration rate was increased in three times ($u = 1.7$ m/hr). In this case $t_c = 359$ hrs and $B_1 = 0.02971°C/hr$ (Table 25-4).

From the last table follows that for large shut-in times the average drilling mud temperature (case when $B_1 = 0$) can be used to estimate the function $\Delta T = f(t_s)$.

Thus it is demonstrated that in deep wells a simple empirical formula approximates the downhole temperature profile during drilling. It is shown that this formula can be combined with an analytical solution and then only one shut-in temperature log is required to estimate the undisturbed (static) formation temperature.

Table 25-4. Values of $\Delta T = T_f - T_s$ at two depths for two wells.

t_s, hrs	Missisippi well $T_f = 187.8°C$, $h = 6,534$ m, $u = 0.566$ m/hr, $t_c = 1200$ hrs, $u = 0.566$ m/hr			Webb County well $T_f = 91.60°C$, $h = 2,003$ m, $u = 0.566$ m/hr, $t_c = 359$ hr, $u = 1.70$ m/hr		
	$R_{in} = 52.6$ $B_o = 163.6°C$ $B_1 = 001783°C/hr$	$t_c = 1077$ hrs $u = 0.566$ m/hr	$t_c = 359$ hrs, $u = 1.70$ m/hr	$R_{in} = 47.91$ $B_o = 68.2°C,$ $B_1 = 0°C/hr$	$R_{in} = 38.46$ $B_o = 63°C$ $B_1 = 0.0291°C/$ hr	$R_{in} = 28.41$ $B_o = 68.2°C$ $B_1 = 0°C/hr$
10	15.68	16.36	13.08	16.48	12.46	15.40
20	13.83	14.40	11.61	14.48	10.81	13.08
30	12.72	13.24	10.74	13.28	9.83	11.70
50	11.32	11.76	9.62	11.76	8.58	9.94
70	10.39	10.77	8.88	10.75	7.75	8.78
100	9.40	9.73	8.10	9.68	6.87	7.55
120	8.89	9.20	7.70	9.14	6.42	6.92
150	8.27	8.54	7.21	8.46	5.87	6.16
200	7.47	7.70	6.57	7.60	5.16	5.22
300	6.35	6.51	5.67	6.38	4.18	4.00
500	4.95	5.04	4.55	4.88	3.04	2.73
600	4.47	4.53	4.15	4.38	2.68	2.35
800	3.74	3.78	3.54	3.63	2.16	1.85
1000	3.22	3.24	3.09	3.09	1.81	1.52
1200	2.82	2.84	2.75	2.70	1.56	1.29

26

The Duration of Temperature Monitoring in Wellbores: Permafrost Regions

Temperature logs are commonly used to determine the permafrost temperature and thickness. When wells are drilled through permafrost, the natural temperature field of the formations (in the vicinity of the borehole) is disturbed and the frozen rocks thaw for some distances from the borehole axis. To determine the static temperature of the formation and permafrost thickness, one must wait for some period after completion of drilling before making geothermal measurements. This is so-called restoration time. Usually a number of temperature logs (3–10) are taken after the well's shut-in. Significant expenses (manpower, transportation) are required to monitor the temperature regime of deep wells. In this chapter we show that in most of the cases when the time of refreezing formations is relatively short (in comparison with the shut-in time) two temperature logs are sufficient to predict formations temperatures during shut-in, to determine the geothermal gradients, and to evaluate the thickness of the permafrost zone. Thus the cost of monitoring the temperature regime of deep wells after shut-in can be drastically reduced. The presence of permafrost has a marked effect on the time required for the near-well-bore formations to recover their static temperatures. The duration of the refreezing of the layer thawed during drilling is greatly dependent on the natural temperature of formation; therefore, the rocks at the bottom of the permafrost refreeze very slowly. A lengthy restoration period of up to ten years or more is required to determine the temperature and thickness of permafrost with sufficient accuracy (Melnikov et al. 1973; Taylor and Judge 1977; Judge et al. 1979, 1981; Taylor et al. 1982; Lachenbruch et al. 1988). Earlier we suggested a "two point method" (Kutasov 1988) which permits one to determine the permafrost thickness from short term (in comparison with the time required for temperature restoration) downhole temperature logs. The "two point method" of predicting the permafrost thickness is based on determining the geothermal gradient in a uniform layer below the permafrost zone. Only temperature measurements for two depths are needed to determine the geothermal gradient. The position of the permafrost base is predicted by the extrapolation of the static formation temperature-depth curve to 0°C. It should be noted that here the permafrost base is defined as the 0°C isotherm. Precise temperature measurements (Taylor and Judge 1977; Judge et al. 1979) taken in 15 deep wells located in Northern Canada (Arctic Islands and Mackenzie Delta) were used to verify the proposed method. Let us assume that at the moment of time $t = t_{ep}$ the phase transitions (water-ice) in formations at a selected depth are completed, i.e., the thermally disturbed formation has frozen. In this case at $t > t_{ep}$ the cooling process is similar to that of temperature recovery in sections of the well below the permafrost base. It is well known (Tsytovich 1975) that the freezing of the water occurs in some temperature interval below 0°C (Figure 14-1, Chapter 14). In practice, however, the moment of time $t = t_{ep}$ (Figure 14-1) cannot be determined. Only shut-in temperatures T_{s1}, T_{s2}, and T_{s3} are measured at a given depth (Figure 14-1). For this case we proposed a method of predicting the formation temperatures (see Chapter 14).

A generalized formula to process field data (for the well sections below and above the permafrost base) was developed. Temperature logs conducted in five wells were used to apply this method (see Chapter 14). It is demonstrated, by using field examples, that for deep boreholes several methods of predicting the permafrost temperature and its thickness should be applied. Low temperatures (from –15°C to –5°C) are typical for upper sections of the well's lithological profile. In this case the refreezing period is short and an empirical formula (Lachenbruch and Brewer 1959) can be utilized to determine permafrost temperature and geothermal gradient. This is not the case for the lower sections of the wellbore where the surrounding formations are at high temperatures (from –3°C to –1°C). Here freezing time is large and only the "three point" method is used. A simple method to process field data (for the well sections below and above the permafrost base) is presented. Temperature logs conducted in three wells were used to demonstrate utilization of this method.

26.1 Shut-in Temperatures–Permafrost Zone

In 1959 in their classical paper Lachenbruch and Brewer (1959) proposed an empirical formula (Eq. (26-1)) to predict the wellbore temperatures during shut-in.

$$T_s = B \ln\left(1 + \frac{t_c}{t_s}\right) + A, \tag{26-1}$$

where A and B are empirical coefficients determined from field measurements.

It is a reasonable assumption that the value of t_c (the "disturbance" period) is a linear function of the depth (h):

$$t_c = t_t\left(1 - \frac{h}{h_t}\right), \tag{26-2}$$

where h_t is the total well depth, t_c is the time of "thermal disturbance" at a given depth, t_t total drilling time, t_s shut-in time. Equation (26-1) was successfully utilized to describe the measured shut-in temperature in the Well 3, Alaska, South Barrow.

The well was drilled for 63 days to a total depth of 2,900 ft. (Lachenbruch and Brewer 1959). U.S. Geological Survey conducted extensive temperature measurements in Alaska (U.S. Geological Survey "Site" File—Alaska, Internet (Boreholes locations 1998)). We selected long term temperature surveys in three wells (Table 26-1) to find out if the proposed Eq. (15-8) (see Chapter 15) can be used when only two temperature logs are available. In Table 26-2 the calculated values of A and B for two wells are presented. For the well Drew Point (Alaska) No.1 we used the values of T_s measured in the first two temperature logs ($t_{s1} = 186$ days, $t_{s2} = 547$ days) to determine empirical coefficients A and B. After this we calculated values of T_s for the last temperature log ($t_s = 2339$ days).

Table 26-1. Input data and location of three wells, Alaska (U.S. Geological Survey "Site" File—Alaska, Internet, 1988).

Site code	PBF	DRP	ESN
Site name	Put River N-1	Drew Point No.1	East Simpson #1
Latitude	70°19'07" N	70°52' 47.14" N	70°55'04.01"N
Longitude	148°54'35" W	153°53'59.93" W	154°37'04.75" W
Surface elevation, m	8	5	4
Casing diameter, cm	51	34	34
Hole depth, m	763	640	600
Date of drill start	02-09-70	13-01-78	19-02-79
Drilling time, days	44	60	51
Number of logs	9	6	5
Shut-in time, days	5-1071	186-2339	155-1947

Similarly, for the well East Simpson #1 we used the values of T_s measured in the first two temperature logs ($t_{s1} = 155$ days, $t_{s2} = 520$ days) to obtain values of A and B. Values of T_s determined for $t_s = 1947$ days (the last temperature log) are also presented in Table 26-2.

Table 26-2. Calculated values of A and B for wells Drew Point #1 and East Simpson #1.

h, m	T_{s1}, °C	T_{s2}, °C	A, °C	T_s, °C	B, °C
		Drew Point #1			
$t_{s1} = 186$ days		$t_{s2} = 547$ days		$t_s = 2339$ days	
50.29	−7.373	−8.519	−9.192	−8.953	6.990
70.10	−6.925	−8.142	−8.854	−8.574	7.639
100.58	−5.892	−7.283	−8.092	−7.774	9.147
120.40	−5.309	−6.638	−7.408	−7.140	9.022
150.88	−4.466	−5.690	−6.394	−6.148	8.751
170.69	−3.951	−5.076	−5.721	−5.518	8.336
199.64	−3.247	−4.199	−4.741	−4.619	7.455
219.46	−2.698	−3.639	−4.173	−4.032	7.672

Table 26-2. contd....

Table 26-2. contd.

h, m	T_{s1}, °C	T_{s2}, °C	A, °C	T_s, °C	B, °C
		East Simpson #1			
t_{s1} = 155 days		t_{s2} = 520 days		t_s = 1947 days	
21.34	−8.019	−9.180	−9.746	−9.601	6.268
42.67	−7.501	−8.817	−9.456	−9.245	7.332
60.96	−7.145	−8.486	−9.135	−8.917	7.683
79.25	−6.699	−7.999	−8.626	−8.437	7.669
100.58	−6.107	−7.427	−8.060	−7.866	8.069
124.97	−5.538	−6.736	−7.308	−7.155	7.644
149.35	−4.935	−6.063	−6.598	−6.474	7.532
173.74	−4.487	−5.499	−5.977	−5.905	7.092
201.17	−3.538	−4.526	−4.989	−4.980	7.338
225.55	−2.775	−3.963	−4.517	−4.308	9.328
249.94	−2.433	−3.275	−3.666	−3.702	7.019

From Eq. (26-1) follows that at $t_s \to \infty$ $T_s \to T_f = A$. Assuming that at values of t_s = 2339 days and t_s = 1947 days the values of T_s are close to the undisturbed temperature of formations, we can see from Table 26-2 that values of T_s and A are in good agreement. In Tables 26-3 and 26-4 we compare the calculated and measured shut-in temperature. The agreement between the values T_{eq} and T_s is very good. US Geological Survey has obtained unique data for the well PUT River, Alaska N-1 (Table 26-5). Indeed, during two month of well's shut-in five temperature logs were taken. For the well Put River N-1 we used values of t_{s1} = 34 days and t_{s2} = 66 days to estimate the empirical coefficients A and B.

Table 26-3. Comparison of measured and observed shut-in temperatures, well East Simpson #1 (t_{s1} = 155 days and t_{s2} = 520 days).

h, m	A, °C	B, °C	T_{eq}, °C	T_s, °C	T_{eq}, °C	T_s, °C	T_{eq}, °C	T_s, °C
			at t_s = 865 days		at t_s = 1608 days		at t_s = 1947 days	
21.34	−9.746	6.268	−9.399	9.344	9.557	-	−9.590	−9.601
42.67	−9.456	7.332	−9.065	−9.060	−9.243	−9.222	−9.280	−9.245
60.96	−9.135	7.683	−8.738	−8.742	−8.919	−8.903	−8.956	−8.917
79.25	−8.626	7.669	−8.243	−8.274	−8.418	−8.440	−8.454	−8.437
100.58	−8.060	8.069	−7.673	−7.692	−7.850	-	−7.886	−7.866
124.97	−7.308	7.644	−6.959	−6.987	−7.118	-	−7.151	−7.155
149.35	−6.598	7.532	−6.272	−6.303	−6.421	-	−6.451	−6.474
173.74	−5.977	7.092	−5.686	−5.734	−5.819	-	−5.846	−5.905
201.17	−4.989	7.338	−4.707	−4.799	−4.836	−4.940	−4.862	−4.980
225.55	−4.517	9.328	−4.180	−4.137	−4.334	−4.285	−4.366	−4.308
249.94	−3.666	7.019	−3.429	−3.524	−3.537	−3.684	−3.560	−3.702
274.32	−3.066	4.944	−2.910	−3.023	−2.982	−3.126	−2.996	−3.161
301.75	−2.113	5.660	−1.950	−2.085	−2.024	-	−2.040	−2.247

Table 26-4. Comparison of measured and observed shut-in temperatures, well Drew Point #1 (t_{s1} = 186 days, t_{s2} = 547 days).

H, m	A, °C	B, °C	T_{eq}, °C	T_s, °C	T_{eq}, °C	T_s, °C	T_{eq}, °C	T_s, °C
			at t_s = 907 days		at t_s = 1259 days		at t_s = 2339 days	
50.29	−9.192	6.990	−8.778	−8.726	−8.892	−8.736	−9.029	−8.953
70.10	−8.854	7.639	−8.417	−8.355	−8.537	−8.292	−8.681	−8.574
100.58	−8.092	9.147	−7.596	−7.470	−7.732	−7.607	−7.896	−7.774
120.40	−7.408	9.022	−6.936	−6.852	−7.066	−6.953	−7.222	−7.140
150.88	−6.394	8.751	−5.962	−5.856	−6.081	−5.961	−6.224	−6.148
170.69	−5.721	8.336	−5.326	−5.233	−5.435	−5.335	−5.566	−5.518
199.64	−4.741	7.455	−4.409	−4.342	−4.500	−4.441	−4.611	−4.619
219.46	−4.173	7.672	−3.847	−3.760	−3.936	−3.860	−4.045	−4.032
243.84	−3.336	5.148	−3.129	−3.043	−3.186	−2.991	−3.255	−3.304

Table 26-5. Measured shut-in temperatures, well Put River N-1.

h, m	Shut-in time, days					
	5	22	34	48	66	1071
30.48	−.400	−2.686	−4.793	−6.252	−7.040	−9.167
45.72	−.300	−2.093	−4.507	−6.012	−6.910	−9.052
60.96	−.250	−2.941	−4.911	−6.148	−6.950	−8.957
91.44	−.300	−1.633	−4.101	−5.646	−6.590	−8.771
121.92	−.210	−.882	−2.565	−4.781	−6.060	−8.520
152.40	−.030	−.976	−1.852	−3.173	−4.760	−8.124
182.88	.020	−.757	−1.217	−2.506	--	−7.619
213.36	.200	−.490	−.805	−1.528	--	−7.144
243.84	.380	−.433	−.608	−.950	−2.660	−6.602
274.32	.640	−.418	−.555	−.823	--	−6.029
304.80	.740	−.379	−.506	−.682	−1.150	−5.462
335.28	.910	−.325	−.451	−.577	--	−4.935
365.76	1.040	−.322	−.452	−.579	−.800	−4.454
396.24	1.230	−.354	−.505	−.644	−.860	−4.039
426.72	1.220	−.280	−.415	−.517	−.630	−3.453
457.20	1.890	−.326	−.395	−.497	--	−2.961
487.68	1.480	−.305	−.398	−.476	--	−2.493
518.16	1.520	−.264	−.309	−.389	--	−2.006
548.64	1.880	−.171	−.316	−.382	--	−1.418
579.12	2.540	−.010	−.236	−.309	−.360	−.778
609.60	2.600	3.491	1.918	1.212	.710	−.195
640.08	6.830	5.275	4.047	3.254	2.640	.761
670.56	9.350	5.948	4.749	4.002	--	1.664
701.04	9.910	7.251	6.022	5.256	4.650	2.885

In Table 26-6 we compare the calculated and measured shut-in temperatures for the well Put River N-1. In this case the difference between the calculated and measured temperatures is significant (Table 26-6). We can conclude that the shut-in times (34 and 66 days) are comparable with the formation freezeback period (Table 26-5).

Table 26-6. Comparison of measured and observed shut-in temperatures, well Put River N-1 (t_{s1} = 34 days, t_{s2} = 66 days).

h, m	A, °C	B, °C	T_{eq}, °C T_s, °C at t_s = 117 days		T_{eq}, °C T_s, °C at t_s = 163 days		T_{eq}, °C T_s, °C at t_s = 1071 days	
30.48	−10.728	6.703	−8.407	−7.970	−8.985	−8.716	−10.432	−9.167
45.72	−10.824	7.265	−8.366	−7.900	−8.979	−8.428	−10.511	−9.052
60.96	−10.246	6.251	−8.180	−7.860	−8.698	−8.263	−9.984	−8.957
91.44	−9.262	5.303	−7.595	−7.620	−8.015	−7.965	−9.052	−8.771
121.92	−10.486	8.121	−8.065	−7.250	−8.678	−7.624	−10.184	−8.520
152.40	−9.235	9.832	−6.467	−6.510	−7.172	−7.026	−8.892	−8.124

For this case we can use the earlier suggested three-point method (Kutasov and Eppelbaum 2003) for predicting the formation temperatures (Table 26-7). Here an additional temperature log (t_s = 48 days) was used. From Table 26-7 follows that the agreement between calculated and measured temperatures is very good.

Table 26-7. Observed (T^*) and calculated (T) shut-in temperatures (°C) at three depths of the Put River N-1 well, Alaska; t_{s1} = 34, t_{s2} = 48, and t_{s3} = 66 days (modified after Kutasov and Eppelbaum 2003).

t_s, days	45.72 m T_f = −9.242°C		60.96 m T_f = −9.075°C		91.44 m T_f = −9.023°C	
	T, °C	T^*, °C	T, °C	T^*, °C	T, °C	T^*, °C
91	−7.546	−7.511	−7.525	−7.497	−7.258	−7.227
117	−7.921	−7.900	−7.867	−7.860	−7.651	−7.620
163	−8.294	−8.428	−8.207	−8.263	−8.040	−7.965
1071	−9.098	−9.052	−8.946	−8.957	−8.875	−8.771

26.2 Temperature Gradient and Estimation of the Permafrost Thickness

In the permafrost areas the rate of heat flow at the frozen-unfrozen interface serves as the main criterion of the steadiness or non-steadiness of the thermal regime. Let us assume that q_t and q_f are the heat flow density at the phase boundary in the unfrozen and frozen zone, respectively.

$$q_f = \lambda_f G_f, \qquad q_u = \lambda_t G_t,$$

where λ_f is the thermal conductivity of frozen formations, λ_t is the thermal conductivity of unfrozen formations, G_f is the geothermal gradient in the frozen zone and G_t is the geothermal gradient in the unfrozen zone.

It is clear that the condition $q_f = q_u$ corresponds to a steady regime, the condition $q_f > q_u$ corresponds to a regime of freezing and the condition $q_f < q_u$ corresponds to a thawing regime. In addition, the change in the heat flow density at the permafrost base (frozen-unfrozen interface) is also an indicator of the climate change in the past (Melnikov et al. 1973). When interpreted with the heat conduction theory, these sources can provide important information of patterns of contemporary climate change. For example, precision measurements in oil wells in the Alaskan Arctic indicate a widespread warming (2–4°C) at the permafrost surface during the 20th century (Lachenbruch et al. 1988). Thus to estimate values of q_u and q_f, it is necessary to determine the geothermal gradient and formation conductivity in the frozen and unfrozen sections of the wellbore. Unfortunately at present no methods are available for *in-situ* determination of formation conductivity. Samples of rocks are usually used to estimate formation thermal conductivity. Experimental studies show that $\lambda_f > \lambda_u$. For a given formation the λ_f/λ_u ratio is mainly a function of total water content (Balobayev 1991). The duration of the refreezing of the layer thawed during drilling

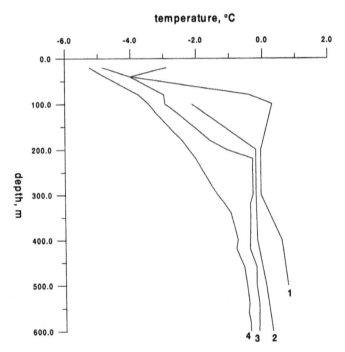

Figure 26-1. Restoration of temperature profile in the Bakhynay borehole 1-R (Melnikov et al. 1973). Temperature surveys 1, 2, 3, and 4 were conducted at shut-in times of 0.4, 1.5, 3.4 and 10.4 years, respectively.

is very dependent on the natural temperature of formation; therefore, the rocks at the bottom of the permafrost refreeze very slowly. In practice the position of the permafrost base (h_p) is estimated by extrapolation. Let us examine the restoration of the natural temperature field by the example of the Bakhynay borehole 1-R (Melnikov et al. 1974).

The borehole was drilled for 23 months (1956–1958) to a depth of 2824 m. Nine temperature logs were performed over a shut-in period of 10 years, but the difference between the temperature of formations and that in the borehole was still greater than the measurement accuracy (0.03–0.05°C). After a shut-in period of 1.5 years the thickness of permafrost was estimated as 470 m instead of 650 m. The restoration of the temperature regime was accompanied by formation of practically zero temperature gradient intervals (Figure 26-1). Therefore, if the shut-in time is insufficient, one may incorrectly attribute the zero temperature gradient intervals to some geological-geographical factors, an example of which in this case is the warming effect of the Lena River (the drilling site is the bank of the river). Below we present several examples of determination of the temperature gradients (G) and evaluation of the permafrost thickness. For each section of the well we used a linear regression program to calculate the coefficients in the following equation

$$T_s = a + gh. \tag{26-3}$$

The position of the transient permafrost base (h_p*) is estimated by extrapolation by using Eq. (26-3) for $T_s = 0°C$.

Therefore:

$$0 = a + gh, \qquad h_p^* = -\frac{a}{g}.$$

$$T_f = b + Gh. \tag{26-4}$$

$$0 = b + Gh, \qquad h_p = -\frac{b}{g},$$

where a and b are constants, h is the depth, h_p is the position of the permafrost base, G is the geothermal temperature gradient, T_s is the shut-in temperature, and T_f is the formation temperature.

As can be seen for the section 21.34 – 249.94 m at t_s = 520 days (Table 26-8) practically g = G (0.0267°C/m). We assumed that at t_s = 1947 days g = G. Similarly, for section 451.10 – 600.46 m at t_s = 520 days the values g and G are very close (0.0408 and 0.0415°C/m).

In Table 26-9 we present an example of data processing for two wells. As can be seen shut-in times of 155 days and 186 days do not enable to estimate the permafrost thickness with a sufficient accuracy.

Table 26-8. The estimated values of permafrost thickness for two sections, East Simpson #1.

t, days	g, G, °C/m	a, b, °C	h_p, h^*_p, m
\multicolumn{4}{c}{Section 21.34 – 249.9 4 m}			
155	0.02501	−8.6303	345.1
520	0.02660	−9.9999	375.9
865	0.02658	−10.236	385.1
1608	0.02750	−10.519	382.1
1947	0.02673	−10.437	390.5
\multicolumn{4}{c}{Section 451.10 – 600.46 m}			
155	0.04008	−13.674	341.2
520	0.04078	−14.492	355.4
865	0.04109	−15.035	365.9
1608	0.04105	−15.139	368.8
1947	0.04150	−15.432	371.8

Table 26-9. An example of data processing for two wells, $R = \dfrac{T_{scal} - T_s}{T_s} \cdot 100\%$.

h, m	T_s, °C	T_{scal} °C	T_s, °C	T_{scal} °C
\multicolumn{2}{c}{East Simpson #1 t_s = 155 d h_z = 340.4 m}		\multicolumn{2}{c}{East Simpson #1 t_s = 520 d h_z = 368.2 m}	
451.1	4.490	4.414	3.906	3.867
475.5	5.348	5.387	4.857	4.883
499.9	6.288	6.359	5.855	5.898
524.3	7.291	7.332	6.915	6.914
551.7	8.509	8.425	8.101	8.056
573.0	9.268	9.276	8.930	8.945
\multicolumn{2}{c}{R_{aver} = 0.99% a_0 = −13.573°C g = 0.03987°C/m}		\multicolumn{2}{c}{R_{aver} = 0.60% a_0 = −14.918°C g = 0.04164°C/m}	
\multicolumn{2}{c}{Drew Point No.1, Well 555 t_s = 186 d, h_z = 266.2 m}		\multicolumn{2}{c}{Drew Point No.1 t_s = 547 d, h_z = 290.3 m}	
400.2	4.740	4.67	3.981	3.925
450.2	6.417	6.408	5.746	5.711
500.2	8.064	8.149	7.381	7.497
550.2	9.829	9.891	9.232	9.284
600.2	11.586	11.632	11.091	11.070
638.9	13.093	12.981	12.510	12.453
\multicolumn{2}{c}{R_{aver} = 0.90% a_0 = −9.2752°C g = 0.03484°C/m}		\multicolumn{2}{c}{R_{aver} = 0.95% a_0 = −10.376°C g = 0.03574°C/m}	

Earlier we developed "two temperature logs method" for determination of the undisturbed formation temperatures. The working formulas are 15-8 and 15-9 (Chapter 15).

Comparing the results of calculation h_p (Table 26-9 at t_s = 520 days and t_s = 547 days) with that (Table 26-11) we can see that the agreement between calculated values of h_p is good. In Table 26-10 we present results of estimation of the formation temperature for two wells. And, finally, we use a linear regression program to determine the geothermal gradient (Table 26-11).

Table 26-10. Calculated formation temperatures by "Two temperature logs method".

h, m	T_{s1}, °C	T_{s2}, °C	T_p °C
Well East Simpson #1			
t_{s1} = 155 days t_{s2} = 520 days			
451.10	4.490	3.906	3.653
475.49	5.348	4.857	4.645
499.87	6.288	5.855	5.669
524.26	7.291	6.915	6.754
551.69	8.509	8.101	7.928
573.03	9.268	8.930	8.788
Drew Point No.1			
t_{s1} = 186 days t_{s2} = 547 days			
400.20	4.740	3.981	3.576
450.19	6.417	5.746	5.391
500.16	8.064	7.381	7.023
550.17	9.829	9.232	8.921
600.15	11.586	11.091	10.837
638.86	13.093	12.510	12.303

26.3 Onset of Formations Freezeback

To plan the schedule of conducting temperature logs it is important to approximately estimate the onset of the formations freezeback. Earlier we introduced the term "safety period"—the length of the shut-in period during which water-base mud remains free from freezing in permafrost areas (Kutasov and Strickland 1988). From physical considerations it is clear that the "safety period" (t_{sp}) can be determined from the condition $T_s(t_{sp}) = 0$°C (Figure 14-1). Thus, the time $t_s = t_{sp}$ can be considered as the onset of the formations freezeback. The magnitude of the "safety period" depends mainly on the duration of the thermal disturbance (drilling time) and on the static temperature of permafrost. Precise temperature measurements (61 logs) conducted by the Geothermal Service of Canada in 32 deep shut-in wells in Northern Canada (Taylor and Judge 1977; Judge et al. 1979, 1981; Taylor et al. 1982) were used to estimate the values of t_{sp} (Kutasov and Strickland 1988). The total drilling time (t_t) for these wells ranged from 4 to 404 days, the total vertical depth (h_t) ranged from 1356 m to 4704 m), and the depth of permafrost (h_p) ranged from 74 m to 726 m. The range of formations temperatures was -0.5°C > T_f > -4.6°C. We have found that the duration of

Table 26-11. Determination of the geothermal gradient, $R = \dfrac{T_{fcal} - T_f}{T_f} \cdot 100\%$.

h, m	T_f, °C	T_{fcal}, °C	R, %
	Well East Simpson #1		
	$R_{aver} = 0.43\%\ h_z = 365.6$ m		
	$T_f = a_0 + Gh$		
	$G = 0.04245$°C/m $a_0 = -15.518$°C		
451.10	3.6530	3.6301	0.63
475.49	4.6450	4.6653	-0.44
499.87	5.6690	5.7002	-0.55
524.26	6.7540	6.7355	0.27
551.69	7.9280	7.8998	0.36
573.03	8.7880	8.8056	-0.20
	Drew Point No.1		
	$R_{aver} = 1.23\%\ h_z = 304.0$ m		
	$G = 0.03652$°C/m $a_0 = -11.103$°C		
400.20	3.5760	3.5131	1.76
450.19	5.3910	5.3388	0.97
500.16	7.0230	7.1638	-2.00
550.17	8.9210	8.9903	-0.78
600.15	10.8370	10.8156	0.20
638.86	12.3030	12.2293	0.60

the "safety period" t_{sp} for a given depth can be approximated with sufficient accuracy as a function of two independent variables: time of thermal disturbance at the given depth (drilling time) and permafrost static temperature (T_f). A regression analysis computer program was used to process field data. It was revealed that the following empirical formula could be used to estimate the safety shut-in period:

$$t_{sp} = 6.12 t_c^{0.817} (-T_f)^{-1.5}, \tag{26-5}$$

where t_c is the thermal disturbance time (in days) at a given depth and temperature is in °C. The value t_d is: $t_d = t_t - t_h$, where t_h is the period of time needed to reach the given depth. The values of t_h can be determined from drilling records. The value t_d can be also estimated from Eq. (26-5). Please note that in our paper (Kutasov and Strickland 1988) a safety factor of 2 was introduced in Eq. (26-5). When planning to conduct a temperature log the condition $t_s \gg t_{sp}$ should be satisfied. We conducted calculations of t_{sp} for the section 335.3 – 548.6 m, well Put River N-1 (Table 26-12).

Table 26-12. The values of t_{sp} for the section 335.3 – 548.6 m, well Put River N-1.

h, m	t_d, days	T_f, °C	t_{sp}, days	$t_s = 5$ days T_s, °C	$t_s = 22$ days T_s, °C
335.3	24.7	−4.93	7.7	.910	−.325
365.8	22.9	−4.45	8.4	1.040	−.322
396.2	21.1	−4.04	9.1	1.230	−.354
426.7	19.4	−3.45	10.8	1.220	−.280
457.2	17.6	−2.96	12.5	1.890	−.326
487.7	15.9	−2.49	14.9	1.480	−.305
518.2	14.1	−2.01	18.7	1.520	−.264
548.6	12.4	−1.42	28.3	1.880	−.171

From Table 26-12 follows that for the temperature logs with $t_s = 5$ days and $t_s = 22$ days the condition $t_s \gg t_{sp}$ is not satisfied. As a result these temperature logs cannot be used for determination of formation temperatures and estimation of the permafrost thickness. It should be remembered that in Eq. (26-5), time is in days and temperature in °C. We can conclude that for large shut-in times the empirical Lachenbruch-Brewer formula can be used with good accuracy to estimate the shut-in temperatures. Only two temperature logs are needed to calculate the coefficients in the Lachenbruch-Brewer formula. Thus two temperature logs enable to predict formations temperatures, to determine the geothermal gradients, and to evaluate the thickness of the permafrost zone. As a result the cost of monitoring the temperature regime of deep wells after shut-in can be drastically reduced (Kutasov and Eppelbaum 2012b). For short shut times (comparable with the time of complete freezeback) we suggest to utilize the "Three Point Method" (Kutasov and Eppelbaum 2003). The approximate evaluation of the onset of formations freezeback will assist in planning the schedule of conducting temperature logs.

27

The Effect of Thermal Convection and Casing on Temperature Regime of Boreholes

After the complete termination of the recovery process of the temperature field (disturbed by drilling) the temperature in the borehole may still differ from the natural temperature of formations; this is due to the distorting influence of free heat convection and of the casing. Because of this, in the interpretation of temperature logs, some investigators assume that small temperature anomalies can be explained by the presence of metallic casing or of convective movements of fluid (gas) in the borehole. At present, no difficulty is anticipated in regard to the establishment of the per se fact of existence of convection in a vertical pipe filled with fluid or gas. Free heat convection arises when the temperature gradient equals or exceeds the so-called critical temperature gradient. The critical temperature gradient (A_{cr}) is expressed through the critical Rayleigh number (Ra_{cr}) and depends on two parameters: on the ratio of formation (casings) and fluid (gas) thermal conductivities (λ_f/λ); on the convective parameter of the fluid. Both parameters are dependent on the temperature (depth). Ostroumov (1952) obtained for a vertical pipe an equation for the determination of the critical Rayleigh number. Taking into account that this publication is not easily accessible to researchers, we present below a numerical solution of this equation (Ostroumov 1952). At determining the value of A_{cr} in most of the field studies it was assumed that thermal conductivity ratio λ_f/λ is infinity (Diment 1967; Gretener 1967; Sammel 1968; Cermak et al. 2007, 2008). This assumption is applicable to metallic casings, but is not valid for uncased wells and boreholes with plastic casings. The variations of the convective parameter with depth can lead (in some sections of the well) to a condition when the temperature gradient is smaller than the critical temperature gradient. In this case the process of free thermal convection cannot be initiated. Below we present this case for an air filled wellbore.

The objective of this chapter is to present results of field studies (and modeling) of thermal convection effects and casings on thermal regime of deep and shallow boreholes.

27.1 Critical Temperature Gradient

As mentioned above, for a model of vertical pipe Ostroumov (1952) obtained an equation for the determination of the critical Rayleigh number Ra_{cr}:

$$kr_0 \left[\frac{J_0(ikr_0)}{-iJ_1(ikr_0)} + \frac{J_0(kr_0)}{J_1(kr_0)} \right] = 2\left(1 - \frac{\lambda}{\lambda_f}\right),$$

where $J_0(x)$, $J_0(ix)$, $J_1(x)$ and $J_1(ix)$ are the Bessel functions of index zero and one,

$$k^4 = \frac{g\beta}{va}(grad\,T)_{cr}.$$

The results of the numerical solution of this equation are presented in Table 27-1 and Figure 27-1.

Table 27-1. The function $Ra_{cr} = f(\lambda_f/\lambda)$ (Ostroumov 1952).

λ_f/λ	Ra_{cr}	λ_f/λ	Ra_{cr}	λ_f/λ	Ra_{cr}
0	67.4	1.02	104.9	6.27	168.0
0.07	70.7	1.59	118.6	12.5	187.4
0.31	81.0	2.43	133.6	58.4	208.5
0.62	92.4	3.78	150.0	∞	215.8

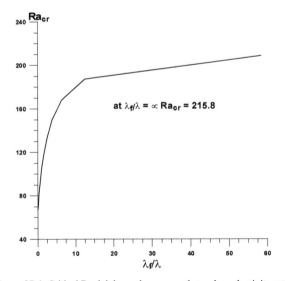

Figure 27-1. Critical Rayleigh number versus thermal conductivity ratio.

Thus, in order to determine A_{cr}, it is necessary to determine the ratio λ/λ_f after which the critical Rayleigh number Ra_{cr} (Table 27-1) and A_{cr} are calculated:

$$Ra_{cr} = \frac{g\beta\beta_{cr}r_0^4}{va}, \quad A_{cr} = Ra_{cr}\frac{va}{g\beta\beta_0^4}, k_p = \frac{g\beta}{va}, \tag{27-1}$$

where λ_f and λ are the thermal conductivities of formations (casing pipes) and of the fluid (gas) in the borehole, respectively, v is the kinematic viscosity, a is the thermal diffusivity of the fluid (gas), β is the coefficient of thermal volumetric expansion, k_p is the convective of thermal volumetric expansion, g is the gravity acceleration, and r_0 is the inside radius of the pipe.

Hales (1937) obtained the following equation for the critical temperature parameter gradient:

$$A_{cr} = \frac{g\beta T}{c_p} + C\frac{va}{g\beta r_0^4}, \tag{27-2}$$

where T is the temperature in K, c_p is the specific heat of fluid at constant pressure, and $C = 216$.

It is easy to see that Hales (1937) does not account the influence of formation thermal properties upon quantity A_{cr}. Hales assumed that thermal conductivity of formations (casing) is infinity (see Table 27-1). Equations (27-1) and (27-2) indicate that the value of A_{cr} strongly depends on the borehole radius. For the ratio of $\lambda_f/\lambda \leq 60$ we found that the values of Ra_{cr} (Table 27-1) can be approximated by the following empirical Eq. (27-3) (Table 27-2), and for $\lambda_f/\lambda > 60$ it can be assumed that $Ra_{cr} \approx 216$.

$$\left. \begin{array}{l} Ra_{cr} = \exp\left(a_0 + a_1 x + a_2 x^2\right), \quad x = \ln\left(1 + \lambda_R\right), \quad \lambda_R = \frac{\lambda_f}{\lambda}, \\ a_0 = 4.2327, \quad a_1 = 0.6350, \quad a_2 = -0.09007 \end{array} \right\} \tag{27-3}$$

Table 27-2. Comparison of "exact" $(Ra_{cr}{}^*)$ and approximated (Eq. (27-3)) values of the critical Rayleigh number Ra_{cr}. $R = (Ra_{cr}{}^* - Ra_{cr}) \cdot 100/Ra_{cr}{}^*$.

λ_R	$Ra_{cr}{}^*$	Ra_{cr}	R
0.	67.4	68.9	−2.2
0.07	70.7	71.9	−1.7
0.31	81.0	81.2	−0.2
0.62	92.4	91.6	−0.9
1.02	104.9	103.0	1.8
1.50	118.6	114.3	3.6
2.43	133.6	131.4	1.6
3.78	150.0	149.2	0.5
6.37	168.0	171.0	1.8
12.5	187.4	195.4	−6.2
58.4	208.5	205.1	1.6

27.2 Convective Parameter, k_p

To conduct calculations after Eq. (27-1) the convective parameter (Eq. (27-4)) should be determined

$$k_p = \frac{g\beta}{va}. \tag{27-4}$$

To estimate the values of k_p and thermal conductivity for water and air as functions of the temperature, the input data were taken from the Internet (The Engineering Toolbox, 20.02.2010), Tables 27-3 and 27-4.

Table 27-3. Input parameters for water (http://www.engineeringtoolbox.com/water-thermal-properties-d_162.html).

T, °C	v, 10^{-6} m²/s	ρ, kg/m³	β, 10^{-3} 1/K	C_p, kJ/kg·K	Pr	λ, W/m·K
10	1.304	1000	0.088	4.193	9.47	0.5774
20	1.004	998	0.207	4.182	7.01	0.5978
30	0.801	996	0.303	4.179	5.43	0.6140
40	0.658	991	0.385	4.179	4.34	0.6279
50	0.553	988	0.457	4.182	3.56	0.6418
60	0.474	980	0.523	4.185	2.99	0.6502
70	0.413	978	0.585	4.190	2.56	0.6611
80	0.365	971	0.643	4.197	2.23	0.6670
90	0.326	962	0.698	4.205	1.96	0.6728
100	0.295	962	0.752	4.216	1.75	0.6837

Table 27-4. Input parameters for air (http://www.engineeringtoolbox.com/air-properties-d_156.html).

T, °C	ρ, kg/m³	c_p, kJ/kg·K	λ, W/m·K	v, 10^{-6} m²/s	β, 10^{-3} 1/K	P_r
0	1.293	1.005	0.0243	13.30	3.67	0.715
20	1.205	1.005	0.0257	15.11	3.43	0.713
40	1.127	1.005	0.0271	16.97	3.20	0.711
60	1.067	1.009	0.0285	18.90	3.00	0.709
80	1.000	1.009	0.0299	20.94	2.83	0.708
100	0.946	1.009	0.0314	23.06	2.68	0.703

The values of thermal conductivity of water were calculated by using the following relationships:

$$a = \frac{v}{Pr}, \qquad \lambda = \rho a c_p,$$

where Pr is the Prandtl number and ρ is the density of water.

The regression analysis programs were used to approximate the convective parameters and thermal conductivities of water and air by simple empirical equations.

For water ($10°C \leq T \leq 100°C$)

$$k_p = a_0 + a_1 T + a_2 T^2,$$

$$a_0 = -0.3183 \cdot 10^{10} \frac{1}{m^3 K}, \quad a_1 = 0.7028 \cdot 10^9 \frac{1}{m^3 K^2}, \quad a_2 = 0.8149 \cdot 10^7 \frac{1}{m^3 K^3} \left.\begin{matrix} \\ \\ \end{matrix}\right\} (27\text{-}5)$$

and the average squared percent of deviation is 0.99.

$$\lambda = \lambda_0 + \lambda_1 T + \lambda_2 T^2,$$

$$\lambda_0 = 0.5602 \frac{W}{m \cdot K}, \quad \lambda_1 = 2.0 \cdot 10^{-3} \frac{W}{m \cdot K^2}, \quad \lambda_2 = -7.9 \cdot 10^{-6} \frac{W}{m \cdot K^3} \left.\begin{matrix} \\ \\ \end{matrix}\right\} (27\text{-}6)$$

and the average squared percent of deviation is 0.28.

For air ($0°C \leq T \leq 100°C$):

$$k_p = a_0 + a_1 T + a_2 T^2,$$

$$a_0 = 0.1438 \cdot 10^9 \frac{1}{m^3 K}, \quad a_1 = -0.2005 \cdot 10^7 \frac{1}{m^3 K^2}, \quad a_2 = 0.9284 \cdot 10^4 \frac{1}{m^3 K^3} \left.\begin{matrix} \\ \\ \end{matrix}\right\} (27\text{-}7)$$

and the average squared percent of deviation is 2.91.

$$\lambda_a = \lambda_0 + \gamma T, \lambda_0 = 0.0242 \frac{W}{m \cdot K}, \quad \gamma = 0.00007 \frac{W}{m \cdot K^2} \left.\begin{matrix} \\ \end{matrix}\right\} (27\text{-}8)$$

and the average squared percent of deviation is 0.2.

To avoid freezing of water or water based drilling muds (in shallow observational wells) in permafrost regions very often oil fluids are used to fill the shallow observational wellbores. The convective parameter of these fluids is very much dependent (due to viscosity) on temperature. For the transformer oil we obtained (Kutasov 1976):

$$k_p = 10^{10} (2.823 - 0.0102T) \exp\left(-\frac{4640}{273 + T}\right) \frac{1}{m^3 10^{-6} °C}, \quad -20°C < T < 20°C. \ (27\text{-}9)$$

The critical temperature gradient decreases with increasing of the convective parameter. The data illustrating this fact are presented in Table 27-5. It was assumed that the well radius is 0.1 m, the thermal conductivity of the formation is $\lambda_f = 2.0$ kcal/ (m·hr·°C) [1 kcal/(m·hr·°C) = 1.163 W/m^{2}K]. The table data (at 20°C) were used to calculate the convective parameter.

Table 27-5. The A_{cr} and fluid conductivity (λ) dependence (after Kutasov and Devyatkin 1973).

Well filler	λ, W/(mK)	λ_l/λ	k, $1/(cm^3 \cdot °C)$	Ra_{cr}	A_{cr}, °C/m
Water	0.586	4	12,500	150	0.00012
Transformer oil	0.0951105	23	4,000	190	0.00048
Air	0.0256	99	100	210	0.0210

27.3 Deep Boreholes

For shallow boreholes (down to 20 m) we have shown that the convective process plays a substantial role in the thermal regime of boreholes (Kutasov and Devyatkin 1964, 1977). Some results of an annual field experiment are presented below. For a long time, efforts by direct temperature observations, failed to establish a quantitative estimate of the effect of free thermal convection on temperature regime of deep boreholes. In the mid-1960s, some results of interesting *in situ* and laboratory experiments (Diment 1967; Gretener 1967; Sammel 1968) were published. It was shown that the convective process basically expresses itself in vertical movements of low periodicity (from several minutes to fifteen minutes and above) and that the extent of these movements does not exceed several diameters of the borehole. Continuous temperature recording attested to the fact that the temperature in the borehole was subjected to oscillations. For a 360 m borehole (diameter 25 cm) filled with water, the amplitude of temperature oscillations (ΔT) did not exceed 0.05°C (Diment 1967). The value of ΔT was approximately proportional to the value of the geothermal gradient (Γ), and for this borehole $\Delta T_{max} = c\Gamma$, $c = 1.25$ m. The Gretener's experiments (1967) on two deep wells in Texas led to analogous results.

As was mentioned by Cermak et al. (2007), the increase in the sensitivity of modern geophysical measuring equipments as well as the enormous data storage capacity of the present-day data loggers has opened a possibility to capture the stochastic features of the temperature variations measured in boreholes. Temperature was monitored in two boreholes in Kamchatka (Russia 2001–2003). Ten-min reading (sampling) interval was selected for the first half-year run followed by a shorter (12 days) experiment with 5-s reading interval. A similar experiment was repeated later in the test borehole Sporilov (Prague, Czech Republic), where four temperature–time series were performed with reading intervals varying from 1 to 20s. All temperature–time series (except the record from the bottom of the hole) displayed intermittent, non-periodic oscillations of temperature of up to several hundredths of degree (Cermak et al. 2007). The experiment in the Sporilov borehole occurred in the internal plastic tube 5 cm in diameter, which is sealed from the influx of ground water in the surrounding strata. In the studied interval (82–130 m) the temperature gradient varies in the range of 0.020–0.0215 K/m. Temperature as a function of time was sampled in 15s intervals and the individual measuring steps took from 1.5 to 2.5 days. The obtained results revealed: (1) temperature–time series present a complex apparently random oscillation pattern with the amplitude of up to 0.045 K, (2) the statistical analysis confirmed a quasi-periodic skeleton of a two-frequency oscillation structure. Shorter periods of 10 upto 30 min are superposed on longer variations with period of several hours (Cermak et al. 2008).

For *in situ* investigations in one or in several wells, it is difficult to encompass a wide range parameters entering into the Rayleigh criterion (number). Because of this, Devyatkin (Kutasov and Devyatkin 1973) selected the thermal modelling approach. Some theoretical assumptions (Ostroymov 1952; Berthold and Börner 2008) indicate that the Rayleigh criterion is the determining criterion in the free thermal convection modelling. Thus, the following relation should be satisfied to:

$$\frac{g\beta A_{cr} r_o^4}{va}\bigg|_m = \frac{g\beta A_{cr} r_o^4}{va}\bigg|_n,$$

where indexes n and m refer to natural (*in situ*) and modelling conditions. Analysis of modelling results indicated that the following function satisfactorily describes the maximum value of the temperature disturbance ($\Delta T_{[ul]max}$):

$$\Delta T_{[ul]max} = \frac{\Gamma}{D(1 - B\log Ra)}, \tag{27-10}$$

where Γ is the vertical temperature gradient, D and B are constants for given filler (Table 27-6).

Table 27-6. Values of coefficients D and B (after Kutasov and Devyatkin 1973).

Borehole filler	Ra	D, 1/m	B
water	1000 – 56300	6.25	0.16
transformer oil	1400 – 16300	10.5	0.21
transformer oil	260 – 1400	35.5	0.30
air	200 – 840	50.5	0.00

Table 27-7 represents the maximal temperature deviations calculated after Eq. (27-10).

Table 27-7. Maximal temperature deviations (°C) at an average temperature 20°C (after Kutasov and Devyatkin 1973).

substance (filling)	grad T, °C/m	well radius, cm			
		4	6	8	10
Water	0.03	0.009	0.012	0.015	0.018
	0.05	0.017	0.022	0.028	-
	0.10	0.040	0.050	-	-
	1.00	0.500	-	-	-
Transformer Oil	0.03	0.003	0.009	0.013	0.020
	0.05	0.007	0.017	0.026	-
	0.10	0.010	0.040	-	-
	1.00	0.600	-	-	-
Air	2.00	0.040	-	-	-

Temperature disturbance due to free thermal convection in deep boreholes usually does not exceed 0.03°C (Table 27-7). It is not difficult to see from Eq. (27-10) that for a small interval of variation of the Rayleigh number, the quantity of $\Delta T_{[ul]max}$ is approximately proportional to the vertical temperature gradient (A). Thus, the data of *in situ* observations (Diment 1967; Gretener 1967; Sammel 1968) and those of modelling are in fair agreement. Let us compare the calculations with field data obtained previously (Kutasov and Devyatkin 1964) for a borehole filled with transformer oil. The radius of the borehole was 4 cm and the vertical temperature gradient at depths of 3 to 5 m attained the value of 2.1°C/m. For an average filler temperature of −2.0°C, the Rayleigh number is 5590. From Eq. (27-10), one finds $\Delta T_{[ul]\ max} = 0.94$°C, while for *in situ* observations the temperature deviation due to convection at depth of 3 m is 1.19°C and at 5 m it is 0.83°C.

In the Hausen borehole three high-resolution temperature logs were measured (Figure 27-2) during 1991–1992. The temperatures are measured using a PT100 sensor: temperature range is between −10 and +150°C, and the relative accuracy is ± 0.001°C. The absolute accuracy, in laboratory conditions is ± 0.03°C, and in borehole conditions is ± 0.05°C, which implies reproductive property of the measurements (Pfister and Rybach 1995).

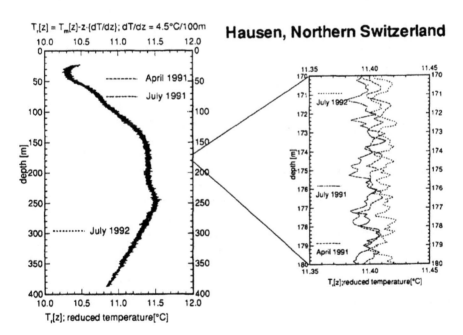

Figure 27-2. Repeated temperature logs of the drillhole in Hausen, Northern Switzerland. On the right, the depth range 170–180 m is magnified (Pfister and Rybach 1995).

The three different temperature logs cannot be separated visually, and the maximum difference between them is 0.05°C. Three main geothermal gradients are visible. The gradient in the depth section 150–250 m is 4.5°C/100 m and was used to reduce the temperature log (Pfister and Rybach 1995). The depth range 170–180 m is magnified for comparison of the three different logs. As was mentioned by Pfister and Rybach (1995) the main features are:

(a) the maximum difference between three logs is 0.04°C,
(b) each log shows variations of the reduced temperature with depth on the amplitude range of 0.02, extending over depth ranges 1–2 m. These variations are assumed to be appearing by the development of convective cells.

Diment and Urban (1983) proposed the following equation (after Pfister and Rybach 1995) for defining the parameters of convection cells

$$R = A\Gamma a, \tag{27-11}$$

where R is the maximum range of temperature variation, A is the cell aspect ratio, Γ is the geothermal gradient, and a is the well radius. The results of calculations after Eq. (27-11) are presented in Table 27-8.

Table 27-8. Comparison of the range of maximum temperature variations in a drillhole due to the presence of small scale convection cells with two different aspect ratios (Pfister and Rybach 1995).

A (aspect ratio) = d/a	3	10
R (range of temperature variation)	0.01°C	0.04°C
Γ (geothermal gradient)	4.5°C/100	4.5°C/100
a (radius of drillhole)	8.75 cm	8.75 cm
D (vertical size of cell)	0.3 m	0.9 m

Field examples show that high-resolution temperature logging can be utilized to obtain information about various groundwater flow regimes (Pfister and Rybach 1995). The authors also concluded that repetitive logs can detect the convective cells and the size of these cells represents a limiting factor when determining a geothermal gradient or modeling the underground thermal regime.

This last conclusion was confirmed by the result of the study conducted by Wisian et al. (1998), where the multiple independent logs demonstrate that most of the "noise" seen in gradient logs (Figure 27-3) is due to convection cells. These cells are predicted to produce temperature fluctuations on the order of 0.001–0.05°C for the gradients in the Smokyhill well. The convection-induced fluctuations would have a 10–50°C/km effect on unfiltered gradients at 1 m intervals (Wisian et al. 1998).

Table 27-9. Maximal temperature deviations, °C (in comparison with the wellbore No. 4) at different depths (after Kutasov and Devyatkin 1964).

H, m	well number		
	1	2	3
1.0	-	6.90	9.80
2.0	0.55	2.35	5.33
3.0	0.65	0.69	1.19
5.0	0.47	0.24	0.83
10.0	0.20	0.42	0.64
15.0	0.15	0.20	0.22

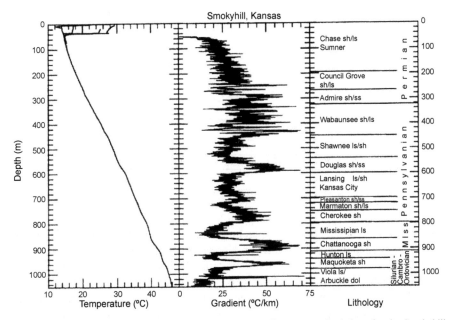

Figure 27-3. Temperature versus depth and temperature gradient versus depth logs for the Smokyhill Kansas well (Wisian et al. 1998)

The effect of convection on the course of the well known in geocriology phenomenon—the time lag of "zero degree" isotherm, is particularly striking. If one assumes that the greater part of the water transforms to ice over the –0.2–0°C temperature interval, the period of transition will vary on basis of the observations of borehole No. 4 (filled with soil) and of borehole No. 2 (covered but not filled). According to the observations in borehole No. 4 (Figure 27-4) this period amounts to around 50 days, and 20 days in borehole No. 2.

Sammel (1968) conducted a series of fields and laboratory tests in wells with diameters of 4.8 and 10.2 cm and showed that water columns are unstable at temperature gradients as low as 0.003°C/m. Magnitudes of thermal oscillations at higher gradients were as large as 0.49°C (Figure 27-4). Large temperature oscillations were observed in 4.8 cm diameter well at all depths to 8.6 m below land surface. For example, at the depth of 2.6 m, where the temperature gradient is 0.66°C/m, the temperature oscillation of 0.275°C occurred during sixteen minute period.

Apparent sizes of convection cells ranged up to 210 cm in height in a 10.2 cm cylinder and to 48 cm in wells 4.8 cm diameter. Comparison of temperatures in adjacent wells showed temperature differences as great as 3°C when thermal gradients were greater than the theoretical critical gradients.

At present monitoring of the temperature regime of shallow wells is widely used to determine the trends in ground surface temperature history (GSTH). In this case accurate subsurface temperature measurements are needed to solve this inverse problem—estimation of the unknown time dependent ground surface temperature (GST). The variations of the GST during the long-term climate changes resulted in

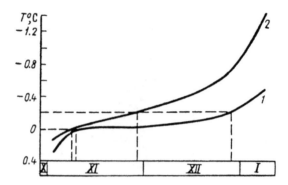

Figure 27-4. Time lag of "zero degree" isotherm in boreholes No. 2 (curve 2) and No. 4 (curve 1) at a depth of 1 m, Yakutsk, Russia (Kutasov and Devyatkin 1964).

disturbance (anomalies) of the temperature field of formations. Thus, the GSTH can be evaluated by analyzing the present precise temperature-depth profiles (e.g., Cermak 1971; Lachenbruch et al. 1988; Beltrami et al. 1992; Shen and Beck 1992; Clauser and Mareschal 1995; Harris and Chapman 1995; Huang et al. 2000; Eppelbaum et al. 2006, 2014). It is obvious that for shallow observational wells the effect of thermal convection on the temperature profiles should be minimized or even eliminated. For this reason the radiuses of these wells should be small and fluids or gases (as well's fillers) with low values of the convective parameters should be utilized.

27.4 Thermal Effect of Casing

Efforts were made to estimate the effect of casing on the distribution of borehole temperatures. Temperature measurements (Kraskovskiy 1934) indicated that the influence of steel casing does not exceed the accuracy of temperature measurements by mercury thermometers ($\pm 0.2°C$). Guyod (1946) has demonstrated, by the means of laboratory modelling, that the presence of a homogeneous steel cylinder having a radius identical to the radius of casing, does not affect the temperature field of the borehole, except at the end portions of the pipes. In order to quantitatively determine this effect, we conducted computer calculations (Kutasov and Devyatkin 1973; Kutasov 1999). The statement of problem was as follows: into a homogeneous medium with thermal conductivity λ and temperature $T_f = T_0 + \Gamma H$, a steel cylinder with thermal conductivity λ_c is introduced. It is required to determine the disturbance of the temperature field due to the cylinder.

The finite-difference method was used to solve the system of Laplace equations, which describe the temperature field inside and outside of the cylinder. The boundary conditions and conditions of coupling, expressed by equality of temperatures and heat flow on the surface of the cylinder are presented in the monograph (Kutasov 1999).

The parameters of the problem were:

$0.03 \leq \Gamma \leq 0.09°C/m$, $0.05 \leq r_c \leq 0.3$ m, $10 \leq \lambda_c/\lambda \leq 30$,

where r_c is the radius.

The processing of the results of numerical calculations (with the accuracy of 0.0001°C) showed that the "disturbance" function for the cylinder's axis could be approximated by the following formula (Kutasov and Devyatkin 1973):

$$T_B(0,H) = \pm 8.3 r_c \Gamma \left(1 - \frac{\lambda}{\lambda_c}\right) \exp\left(-\frac{0.12}{r_c}|H_{1,2} - H|\right). \tag{27-12}$$

Here H_1 and H_2 are the vertical coordinates of the top and bottom of casing, respectively. The plus sign is attributed to H_1 and the minus sign to H_2. It should be noted that, since the casing pipes are replaced by a solid cylinder (as did Guyod (1946)), the influence of the casing on the borehole temperatures would be actually smaller than $T_B(0, H)$. Eq. (27-12) shows that the function $T_B(0, H)$ attains a maximum at the ends of the casing pipe. Usually, $\lambda_c \gg \lambda$ and for this reason, in practice, the value of the "disturbance" function will practically be determined by the radius of the casing, and by geothermal gradient. As an illustration, consider the computer results for one variant of the problem (Figure 27-5).

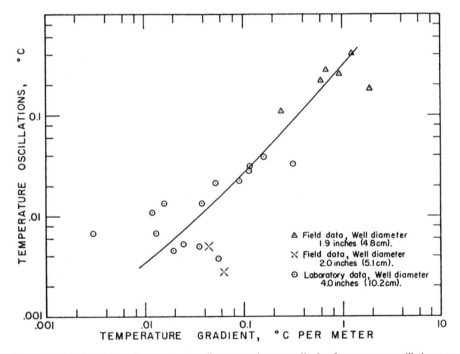

Figure 27-5. Relationship of temperature gradient at a point to amplitude of temperature oscillations at that point (Sammel 1968).

One can see that the maximum value $T_B(0, H_2) = 0.023$°C, while at a distance of 1 m from H_2, the quantity T_B is already down to 0.007°C. Therefore one realizes that the distorting effect due to casing pipes is small and its influence is localized to the ends of the pipes, and is independent of time. However, the value of distortion of the

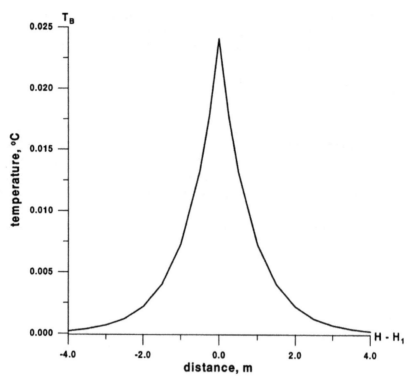

Figure 27-6. The temperature disturbance along the axis of the cylinder; $r_c = 0.1$ m, $\lambda/\lambda_c = 1/30$, $\Gamma = 0.03°C/m$ (after Kutasov and Devyatkin 1973).

natural temperature field, due to convective movements at a given depth, does vary with time and, during a certain period of time, attains its amplitude value ΔT_{max}. Thus the latter quantity, in the final analysis, determines the limiting point in the accuracy that can be reached at temperature surveys in deep boreholes.

The field and modeling results indicate that for deep wells the extent of distortion of the natural temperature field, due to thermal convection mainly depends on the geothermal gradient, varies with time and attains its amplitude value of ΔT_{max} = 0.01–0.05°C. Thus, this temperature range determines the limiting point in the accuracy that can be reached at temperature surveys in deep boreholes (Eppelbaum and Kutasov 2011).

The results of temperature observations in shallow wells show that due to high temperature gradients the effect of thermal convection can be a significant one. However, when designing shallow observational wells the effect of thermal convection can be minimized or even eliminated. This can be achieved by drilling wells with small diameters and utilizing fluids or gases (as well's fillers) with low values of the convective parameters. In this case the critical gradient of temperature will exceed the natural temperature gradient.

The distorting effect due to casing pipes is small and its influence is localized to the ends of the pipes.

APPENDIX A
Hydrostatic Pressure

Let us consider a simple case when the formation pore pressure is equal to the hydrostatic pressure of the formation fluid. The driving force at pressure well tests is the differential pressure—bottom-hole mud pressure minus the pore pressure. For high temperature, high pressure (HTHP) wells the hydrostatic pressure may significantly affect the differential pressure. Below we present methods of predicting density of drilling fluids, water and brines at HTHP. Equations which can be utilized to estimate the hydrostatic pressure are also presented.

A1 Fluid Densities at High Temperatures and High Pressures (HTHP)

At present, a material balance compositional model is used to predict the density of drilling muds and brines at downhole conditions (Hoberock et al. 1982; Sorelle et al. 1982). To use this method, only laboratory density measurements for oils and brines at elevated temperatures and pressures are needed because, conventionally the compressibility and thermal expansion of the solid components are assumed to be very small, and therefore can be neglected. But if a significant amount of chemicals are present in the mud, some chemical interaction can cause changes in the solid-fluid system. It is known, for example, that water-based muds start to degenerate at elevated temperatures. In these cases the compositional model can not be used and laboratory density measurements of water base muds and brines at elevated temperatures and pressures are needed. Earlier we had suggested an empirical equation of state for drilling muds and brines (Kutasov 1988, 1989a). This simple formula will allow one to predict the density of water/oil base muds and brines at downhole conditions. A minimum of input data is required to calculate the coefficients in this formula. Application of the suggested formula may reduce the time and cost of laboratory density tests.

In physics, pressures encountered in deep wells (up to 30,000 psia) are considered only as moderate pressures. Within this range of pressures, the coefficient of isothermal compressibility is a weak function of pressure and can be assumed as a constant for many fluids. Our analysis of laboratory density test data (Hubbard 1984; McMordie et al. 1982; Potter and Brown 1977) for water and oil base muds and brines has shown that their coefficient of thermal (volumetric) expansion can be expressed as a linear function of temperature and the coefficient of isothermal compressibility is practically a constant. We have found that the following empirical formula can be used as an equation of state for either water or oil-based drilling muds, or for brines:

$$\rho = \rho_0 \exp\left[\alpha p + \beta(T - T_s) + \gamma(T - T_s)^2\right]$$
(A-1)

or

$$\ln \rho = \ln \rho_0 + \alpha p + \beta(T - T_s) + \gamma(T - T_s)^2,$$

where T is the temperature (°F), p is the pressure (psig), $T_s = 59°F = 15°C$ (International standard temperature), ρ is the fluid density (ppg); ρ_0 is the fluid density (ppg) at standard conditions ($p = 0$ psig, $T = 59°F$); α (isothermal compressibility), β and γ are constants. A regression analysis computer program was used to process density-pressure-temperature data (Kutasov 1999) and to provide the coefficients of Eq. (A-1). Several investigators (Babu 1996; Kårstad and Aadnøy 1998; Carcione and Poletto 2000) believe that this equation represents the measured data more accurately than other models for majority of drilling muds and brines. From Eq. (A-1) the coefficient of isobaric thermal expansion is (taking into account that $V = 1/\rho$):

$$\beta_v = \frac{1}{V}\left(\frac{\partial V}{\partial T}\right)_p = -\beta - 2\gamma(T - T_s).$$
(A-2)

A2 Water Formation Volume Factor

In water drive reservoirs large quantities of water production may be required to obtain maximum oil recoveries. To maintain the reservoir pressure a significant amount of water is injected into producing formation. In many oil fields the number of injection wells can be close to the number of production wells. Knowledge of the water formation volume factor (B_w) is needed in material balance calculations to predict the change in water volume that occurs between the surface and reservoir.

The value of B_w is defined as water volume under reservoir conditions (reservoir barrels) is divided by water volume under standard conditions. The volume under standard surface conditions is expressed in stock tank barrels (STB) and under reservoir conditions in reservoir barrels (RB). B_w value is a function of temperature and pressure (p). When the thermal expansion of water is compensated by the compression due to the high reservoir pressure, then $B_w = 1.00$ RB/STB.

Below an equation is suggested which will allow one to calculate the values of B_w for temperatures up to 390°F (200°C) and pressures up to 26,000 psia (1,800 bar). Our analysis of laboratory density test data (Burnham et al. 1969) shows that for pure water the coefficient of thermal (volumetric) expansion can be expressed as a linear function of temperature, and the coefficient of isothermal compressibility is practically a constant. We have determined that the coefficients of empirical Eq. (A-1) can be also used as an equation of state (pressure-density-temperature dependence) for pure water (Kutasov 1989b) (Table A-1).

Table A-1. Coefficients in Eq. (A-1) for water (Kutasov 1989b).

Temperature interval, °C (°F)	Pressure interval, bar (psig)	No. of points	p_0, ppg	α, 1/pig ·10^{-6}	β, 1/°F ·10^{-4}	γ, (1/°F)² ·10^{-7}	$\Delta p/p$, ·100%
20–100 (68–212)	100–1,100 (1,436–15,939)	36	8.3555	2.7384	−1.5353	−7.4690	0.039
100–200 (212–392)	300–1,800 (4,336–26,092)	76	8.4172	3.1403	−2.8348	−3.7549	0.175
20–200 (68–392)	100-1,800 (1,436–26,092)	103	8.3619	3.0997	−2.2139	−5.0123	0.172

Results also show that for temperatures up to 100°C Eq. (A-1) is very accurate because, for these temperatures, more precise laboratory test data were available. Taking into account that the specific volume (v) is defined as $v = 1/\rho$, we obtained a relationship for the water formation volume factor

$$B_w = \frac{\rho^*}{\rho}\exp\left[-\alpha p - \beta\left(T - T_s\right) - \gamma\left(T - T_s\right)^2\right],$$

(A-3)

where $\rho^* = 8.3380\ ppg = 0.9991$ g/cm³ is the density of water under standard surface conditions ($T = 59°F$, $p = 0$ psig).

A3 Equation of State for Sodium Chloride Brine at HTHP

Solids free brines are frequently used as completion or workover fluids. In practice the halide brines are most commonly used because they usually represent the lowest cost for a given density (Stephens and Lau 1998). The density of brines must be sufficient to produce a needed overbalance (usually about 200 psi) above the bottom-hole reservoir pressure. In high pressure and high temperature (HPHT) wells the densities of brines can be significantly different from those measured at surface conditions. Calculations have shown that bottom-hole pressures predicted with constant surface densities to be in error by hundreds of psi. In this note we present an equation of state (pressure-density-temperature dependence) for sodium chloride brine. The empirical coefficients in this equation are expressed as functions of the brine weight concentration (W). Thus for an actual temperature profile and specified downhole pressure the needed brine concentration can be determined. Knowledge of pressure exerted by brine as function of downhole temperature, vertical depth, and brine concentration in HPHT wells will allow one to control the overbalance levels.

Density: Our analysis of laboratory density test data (Potter and Brown 1977) for sodium chloride brines has shown that their coefficient of thermal (volumetric) expansion can be expressed as a linear function of temperature and the coefficient of isothermal compressibility is practically a constant. It was found that an empirical formula (Eq. (A-1)) can be used as an equation of state for *NaCl* brines (Table A-2).

Table A-2. Measured, ρ^* (Potter and Brown 1977) and predicted, ρ densities for two sodium chloride solutions (Kutasov 1991).

p, psig	T, °F	$W = 5\%$			$W = 25\%$		
		ρ^*, ppg	ρ, ppg	$\rho-\rho^*$, ppg	ρ^*, ppg	ρ, ppg	$\rho-\rho^*$, ppg
8688	167	8.696	8.692	−.004	9.889	9.897	0.008
11588	167	8.763	8.792	0.029	9.964	9.985	0.021
8688	212	8.604	8.584	−0.021	9.789	9.783	−0.007
11588	212	8.679	8.682	0.003	9.873	9.869	−0.003
14489	212	8.738	8.782	0.044	9.931	9.957	0.026
11588	257	8.579	8.559	−0.021	9.772	9.750	−0.023
14489	257	8.637	8.657	0.019	9.831	9.836	0.005
18115	257	8.704	8.782	0.077	9.898	9.946	0.048
14489	302	8.521	8.518	−0.003	9.714	9.711	−0.003
18115	302	8.596	8.641	0.045	9.798	9.819	0.021
21741	302	8.688	8.765	0.077	9.881	9.928	0.047
14489	347	8.379	8.366	−0.013	9.597	9.581	−0.016
18115	347	8.462	8.486	0.024	9.681	9.688	0.007
21741	347	8.562	8.608	0.046	9.772	9.796	0.023
18115	392	8.312	8.320	0.007	9.564	9.553	−0.011
21741	392	8.454	8.439	−0.015	9.656	9.659	0.004
25367	392	8.537	8.561	0.023	9.739	9.767	0.028
18115	437	8.153	8.141	−0.012	9.430	9.414	−0.016
21741	437	8.270	8.258	−0.012	9.530	9.519	−0.012
25367	437	8.404	8.377	−0.027	9.631	9.625	−0.006
21741	482	8.103	8.066	−0.037	9.405	9.375	−0.030
25367	482	8.237	8.182	−0.054	9.505	9.479	−0.026
28993	482	8.362	8.300	−0.062	9.606	9.585	−0.021

A multiple regression analysis computer program was used to process laboratory density-pressure-temperature data and to provide the coefficients of Eq. (A-1) (Table A-3). It was also found (Kutasov and Seman 2001) that the coefficients of Eq. (A-1) (Table A-3) can be expressed (with good accuracy) as second degree polynomials (Eqs. (A-4)–(A-7), Table A-4)). Thus now the value of brine concentration is directly incorporated into equation of state.

$$\rho_0 = \rho_{00} + \rho_{01}W + \rho_{02}W^2, \tag{A-4}$$

$$\alpha = \alpha_0 + \alpha_1 W + \alpha_2 W^2, \tag{A-5}$$

$$\beta = \beta_0 + \alpha_1 W + \beta_2 W^2, \tag{A-6}$$

$$\gamma = \gamma_0 + \gamma_1 W + \gamma_2 W^2, \tag{A-7}$$

where W is the weight concentration of the sodium chloride brine.

Table A-3. Coefficients in Eq. (A-1) for *NaCl* brines (Kutasov 1999).

W, %	ρ_0, ppg	α, 1/psig·10^{-6}	$-\beta$, 1/°F ·10^{-4}	$-\gamma$, 1/°F²·10^{-7}	$\Delta\rho/\rho$·100, %
1	8.3562	4.3447	1.8070	4.90363	0.41
3	8.4742	4.1310	1.7128	4.67334	0.37
5	8.5908	3.9414	1.6008	4.52541	0.34
7	8.7128	3.7431	1.5436	4.28340	0.34
9	8.8350	3.5928	1.5105	4.05636	0.34
11	8.9624	3.4504	1.5421	3.71040	0.35
13	9.0870	3.3187	1.5505	3.43119	0.35
15	9.2152	3.2549	1.5682	3.21697	0.34
17	9.3472	3.1678	1.6837	2.84163	0.33
19	9.4779	3.1076	1.7719	2.52752	0.20
21	9.6242	3.0667	1.8977	2.18582	0.27
23	9.7504	3.0466	2.0438	1.82556	0.24
25	9.8865	3.0519	2.1967	1.48402	0.21

Table A-4. Coefficients in Eqs. (A-4)–(A-7).

ρ_0	α	β	γ
	Sodium chloride brine		
ρ_{00} = 8.2978 ppg	α_0 = 4.4644·10^{-6} 1/psig	β_0 = −1.8658·10^{-4} 1/°F	γ_0 = −5.0055·10^{-7} 1/(°F)²
ρ_{01} = 5.7636·10^{-2} ppg/W	α_1 = −1.1953·10^{-7} 1/psig·W	β_1 = 6.6435·10^{-6} 1/°F·W	γ_1 = 9.2761·10^{-9} 1/(°F)²·W
ρ_{02} = 2.419·10^{-4} ppg/W^2	α_2 = 2.5231·10^{-9} 1/psig·W^2	β_2 = −3.2144·10^{-7} 1/°F·W^2	γ_2 = 1.9591·10^{-10} 1/(°F·W)²
	Calcium chloride brine		
ρ_{01} = 8.2526 ppg	α_0 = 2.7322·10^{-6} 1/psig	β_0 = −2.5751·10^{-4} 1/°F	γ_0 = −6.8779·10^{-7} 1/(°F)²
ρ_{02} = 6.5577·10^{-2} ppg/W	α_1 = −4.5785·10^{-8} 1/psig·W	β_1 = 1.6008·10^{-6} 1/°F·W	γ_1 = 2.1628·10^{-8} 1/(°F)²·W
ρ_{03} = 3.6570·10^{-4} ppg/W^2	α_2 = 4.0103·10^{-10} 1/psig·W^2	β_2 = −4.2663·10^{-8} 1/°F·W^2	γ_2 = 9.1364·10^{-12} 1/(°F·W)²

A4 Density of Calcium Chloride Brine at HTHP

In this section we present an equation of state (pressure-density-temperature dependence) for calcium chloride brine. The empirical coefficients in this equation are expressed as functions of the brine weight concentration (W). It was also shown earlier that for sodium chloride brines the empirical coefficients in Eq. (A-1) can be expressed as quadratic functions of the brine weight concentration (Kutasov and Seman 2001).

Thus for an actual temperature profile and specified downhole pressure the needed brine concentration can be determined. Knowledge of pressure exerted by brine as function of downhole temperature, vertical depth, and brine concentration in HPHT wells will allow one to control the overbalance levels.

The $CaCl_2$ brine concentration (Gates and Wood 1985, 1989) was expressed in molalities and the density differences, $\Delta\rho = \rho - \rho_w$ were measured (ρ is the density of the solution and ρ_w is the density of water at the same pressure and temperature). Temperature interval used in our study is 77–350 °F and the pressure range is 0–5900 psig (0.1013–40 MPa). The molalities were converted to weight concentrations and the values of $\rho = \Delta\rho + \rho_w$ were calculated by using Eq. (A-1) for the determination of water density (Kutasov 1989b). The coefficients in Eq. (A-1) for water are presented in Table A-1. In this case for the convenience of processing the laboratory data we selected that $T_s = 122°F$ (50°C) and in Eq. (A-1) ρ_0 is the fluid density (ppg) at $p = 0$ psig and $T_s = 122°F$. We should note that usually $T_s = 59°F = 15°C$, the International standard temperature. A multiple regression analysis computer program was used to process laboratory density-pressure-temperature data and to provide the coefficients of Eq. (A-1) (Table A-5). Our analysis of laboratory density test data (Gates and Wood 1985, 1989) for calcium chloride brines has shown that their coefficient of thermal (volumetric) expansion can be expressed as a linear function of temperature and the coefficient of isothermal compressibility is practically a constant (Kutasov 2007). It was also found that the coefficients of Eq. (A-1) (Table A-5) can be expressed (with good accuracy) as second degree polynomials (Eqs. (A-4)–(A-7), Table A-4). Thus now the value of brine concentration is directly incorporated into equation of state. The coefficient of thermal expansion depends on the temperature and weight concentration (Eq. (A-2), and Figure A-1). In Tables A-6 and A-7 the measured ($\rho*$)

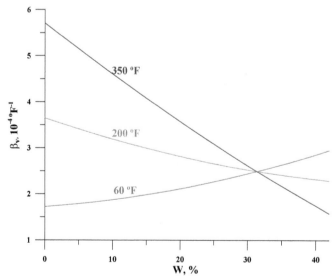

Figure A-1. Coefficient of thermal (volumetric) expansion versus calcium chloride brine concentration.

and calculated (ρ) values (Eq. (A-1)) of fluid density are compared. The results show a good agreement between the measured and predicted densities.

Table A-5. Coefficients in Eq. (A-1) for $CaCl_2$ brines. $T_s = 122°F$ (50°C).

W, %	ρ, ppg	$\alpha \cdot 10^{-6}$, 1/psig	$-\beta \cdot 10^{-4}$, 1/°F	$\gamma \cdot 10^{-7}$, $1/(°F)^2$
2	8.385	2.6422	2.5447	−6.4461
4	8.521	2.5555	2.5179	−6.0124
6	8.659	2.4719	2.4944	−5.5780
8	8.801	2.3916	2.4743	−5.1429
10	8.945	2.3145	2.4576	−4.7071
12	9.092	2.2405	2.4444	−4.2705
14	9.242	2.1698	2.4346	−3.8332
16	9.395	2.1023	2.4281	−3.3951
18	9.551	2.0380	2.4251	−2.9564
22	9.872	1.9190	2.4294	−2.0766
24	10.037	1.8644	2.4366	−1.6356
26	10.205	1.8129	2.4473	−1.1939
28	10.375	1.7646	2.4613	−0.7515
30	10.549	1.7196	2.4788	−0.3084
32	10.726	1.6777	2.4997	0.1355
34	10.905	1.6391	2.5240	0.5802
36	11.087	1.6037	2.5517	1.0255
38	11.273	1.5715	2.5828	1.4716
40	11.461	1.5424	2.6174	1.9184

Table A-6. Measured (ρ^*) (Thomas et al. 1984) and predicted (ρ) densities of $CaCl_2$ brines, $p = 0$ psig.

W, %	ρ^*, g/cm^3	ρ, g/cm^3	$(\rho^*-\rho)/\rho^* \cdot 100$, %	ρ^*, g/cm^3	ρ, g/cm^3	$(\rho^*-\rho)/\rho^* \cdot 100$, %
		25°C			100°C	
2	1.0135	1.0150	−0.15	0.9748	0.9769	−0.22
4	1.0302	1.0314	−0.12	0.9915	0.9933	−0.18
6	1.0471	1.0481	-0.10	1.0085	1.0100	−0.15
8	1.0643	1.0652	−0.09	1.0257	1.0270	−0.13
10	1.0818	1.0827	−0.09	1.0432	1.0444	−0.12
12	1.0997	1.1006	−0.08	1.0610	1.0621	−0.10
14	1.1180	1.1188	−0.07	1.0790	1.0801	−0.10
16	1.1366	1.1374	−0.07	1.0973	1.0985	−0.11
18	1.1557	1.1564	−0.06	1.1160	1.1171	−0.10
20	1.1753	1.1757	−0.04	1.1352	1.1361	−0.08
25	1.2260	1.2258	0.02	1.1846	1.1850	−0.03
30	1.2790	1.2782	0.07	1.2359	1.2359	0.00
35	1.3345	1.3329	0.12	1.2893	1.2887	0.05
40	1.3927	1.3901	0.19	1.3450	1.3434	0.12

Table A-7. Measured (ρ^*) and calculated (ρ) densities of CaCl$_2$ brines.

W, %	Gates and Wood 1985				
	p, MPa	T, °C	ρ, g/cm^3	ρ^*, g/cm^3	$\Delta\rho/\rho^*100$, %
35.60	10.38	25	1.3429	1.3445	0.12
35.60	17.13	25	1.3450	1.3460	0.07
35.60	31.29	25	1.3495	1.3497	0.02
35.60	40.71	25	1.3525	1.3532	0.05
25.08	10.38	25	1.2300	1.2306	0.04
25.08	17.13	25	1.2323	1.2325	0.02
25.08	31.30	25	1.2369	1.2370	0.01
25.08	40.51	25	1.2400	1.2402	0.02
18.21	10.38	25	1.1620	1.1622	0.02
18.21	17.13	25	1.1643	1.1643	0.00
18.21	31.20	25	1.1691	1.1689	−0.02
18.21	40.30	25	1.1722	1.1722	0.00
10.04	10.38	25	1.0868	1.0870	0.02
10.04	17.13	25	1.0893	1.0893	0.00
10.04	31.20	25	1.0944	1.0944	0.00
10.04	40.64	25	1.0979	1.0979	0.00
5.28	10.38	25	1.0460	1.0461	0.01
5.28	17.13	25	1.0486	1.0486	0.00
	Wimby and Berntsson 1994				
34.71	0.10	18.40	1.3339	1.3333	−0.05
34.71	0.10	68.83	1.3033	1.3030	−0.02
39.58	0.10	19.76	1.3889	1.3882	−0.05
39.58	0.10	69.91	1.3562	1.3557	−0.03
42.70	0.10	19.82	1.4259	1.4238	−0.15
42.70	0.10	69.99	1.3914	1.3903	−0.08

A5 Hydrostatic Pressure—a General Equation

A5.1 Equivalent static density

For a static well the pressure at any well's depth is equal to the hydrostatic pressure exerted by the column of the drilling fluid. During drilling fluid circulation an additional pressure drop is required to overcome friction forces opposing the flow of fluid in the annulus. The total pressure during circulation is expressed through the equivalent circulating density. For a successful drilling the mud weight should conform to two conditions: the static mud weight must be able to control the formation pressures and to provide sufficient support to prevent hole collapse, and the equivalent circulating density during circulation should not exceed the fracture gradient.

The mud density increases with pressure and reduces with the increase of the temperature. Many drilling operators consider this compensating effect as a basis for using the surface mud density for calculation of the hydrostatic pressure. In oilfield units (ppg, psig, ft) the following relationship is usually used

$$p^* = B_c \rho_0 h, \qquad B_c = 0.052 \frac{\text{psi}}{\text{ppg} \cdot \text{ft}}, \tag{A-8}$$

where h is the vertical depth, p^* is the pressure at constant density ρ_0, g is the gravity acceleration, and B_c is the conversion factor.

In practice, pressure and pressure gradients are conveniently expressed in density units. Therefore, it is helpful to introduce an equivalent static density (ESD) variable which incorporates the equation of state (pressure-density-temperature dependence) of the drilling mud, well depth and downhole temperatures. This will allow one to evaluate the effect of the vertical depth and downhole temperature on hydrostatic pressure. Indeed, the difference between the surface mud density and the depth dependent values of ESD will show to what extent the downhole conditions may affect the drilling mud program design for high pressure high temperature (HPHT) wells. In oilfield units the equivalent static density is defined by

$$ESD = \frac{p}{B_c D}, \tag{A-9}$$

where D is the vertical depth (ft).

Below we present a general formula for calculating the downhole hydrostatic fluid (mud) pressure. It will be shown that in many cases the effect of the temperature and pressure on drilling mud density should be taken into account at downhole mud pressure predictions.

From physics it is known that

$$dp = \rho g \, dh, \tag{A-10}$$

where dp is the increment given to pressure, $\rho(T,p)$ is the mud density, g is the gravity acceleration, and dh is the increment given to the vertical depth. As was shown earlier the empirical formula (Eq. (A-1)) can be used as an equation of state for either water or oil-base drilling muds, or for brines:

Field and analytical investigations (Kutasov 1999) have shown that the downhole circulating temperatures can be approximated by a linear function of depth

$$T = a_0 + a_1 h. \tag{A-11}$$

From this formula it follows that the stabilized outlet (at $h = 0$) fluid temperature is $T_{out} = a_0$ and the stabilized downhole mud temperature is

$$T_{bot} = T_{out} + a_1 H, \qquad a_1 = \frac{T_{bot} - T_{out}}{H} \tag{A-12}$$

Therefore, to calculate the coefficients, a_0 and a_1 the values of the outlet (flowline) and the bottom-hole circulating mud temperature should be known. The outlet mud temperature is routinely recorded at mud density logging. The bottom-hole circulating temperature can be approximated with sufficient accuracy by an empirical equation as a function of two independent variables: the geothermal gradient and the bottom-hole static (undisturbed) temperature (see Eq. (A-13)).

It was found that the bottom-hole circulating temperature (T_{mb}) can be approximated with sufficient accuracy as a function of two independent variables: the geothermal gradient, Γ and the bottom-hole static (undisturbed) temperature T_{fb}.

$$T_{bot} = d_1 + d_2\Gamma + (d_3 - d_4\Gamma)T_{fb}. \tag{A-13}$$

For 79 field measurements (Kutasov and Targhi 1987), a multiple regression analysis computer program was used to obtain the coefficients of formula.

$d_1 = -50.64°C$ ($-102.1°F$), $d_2 = 804.9$ m (3354 ft),
$d_3 = 1.342$, $d_4 = 12.22$ m/°C (22.28 ft/°F).

These coefficients are obtained for

$74.4°C$ ($166°F$) $\leq T_{fb} \leq 212.2°C$ ($414°F$),
$1.51°C/100$ m ($0.83°F/100$ ft) $\leq \Gamma \leq 4.45°C/100$ m ($2.44°F/100$ ft).

Therefore, Eq. (A-13) should be used with caution for extrapolated values of T_{fb} and Γ. The accuracy of the results (Eq. (A-13)) is 4.6°C, and was estimated from the sum of squared residuals.

For oil-based muds the coefficient γ mainly has a positive sign. The values of αp and γx^2 are very small and we can assume that

$$\exp\left[\gamma(T - T_s)\right] \approx 1 + \gamma(T - T_s)^2, \qquad \exp(-\alpha p) \approx 1 - \alpha p + \frac{(\alpha p)^2}{2}. \tag{A-14}$$

Taking into account Eqs. (A-1), (A-11) and (A-14), the integration of Eq. (A-10), yields (Kutasov 1999).

$$p = \frac{1}{\alpha} - \sqrt{\frac{1}{\alpha^2} - \frac{2F}{\alpha}}, \tag{A-15}$$

where

$$F = B_0 (B_1 B_2 - A_0), \tag{A-16}$$

$$A_0 = 1 + \gamma\left[(a_0 - T_s)^2 - \frac{2}{\beta}(a_0 - T_s) + \frac{2}{\beta^2}\right], \tag{A-17}$$

$$B_0 = \frac{\rho_0 g}{a_1 \beta}\exp[\beta(a_0 - T_s)], \tag{A-18}$$

$$B_1 = \exp(\beta\beta_1 h), \tag{A-19}$$

$$B_2 = 1 + \gamma\left[(a_0 + a_1 h - T_s)^2 - \frac{2}{\beta}(a_0 + a_1 h - T_s) + \frac{2}{\beta^2}\right]. \tag{A-20}$$

A5.2 Example

A major oil company took extensive circulating temperature data while drilling and completing a deep Mississippi well (Wooley et al. 1984). For the depth of 7214 m (23669 ft) the measured values of the stabilized outlet (flowline) and bottom-hole mud temperatures were (Kutasov 1999):

$$T_{out} = 55.6°C \ (132°F) \ \text{and} \ T_{bot} = 172.2°C \ (351°F).$$

The geothermal temperature and mud circulating temperature profiles are presented in Figure A-2. For comparison we conducted calculations assuming that both water-base (WBM, No. 6) and oil-base muds (OBM, No. 12) were considered in the designing of the drilling mud program (McMordie et al. 1982). The results of calculations after Eqs. (A-1) and (A-15) are presented in Table A-8 and Figure A-3. We also compared values of hydrostatic pressure (Eq. (A-15)) with those ($p*$) calculated by conventional constant-surface-mud-density method (Eq. (A-8)).

As follows from Table A-7 the effect of the temperature and pressure on drilling mud density should be taken into account at downhole mud pressure predictions.

Let us assume that the well was shut in for a long time and the wellbore temperature is close to the static (undisturbed) temperature of formations (Table A-8). The hydrostatic pressure for this case (p_s) is presented in Table A-7. One can observe that for the 3,000–7,200 m section of the well the density of water-based mud is practically constant, while for the oil-base mud its value markedly increases with depth (Table A-8).

Table A-8. Calculated hydrostatic pressure. Mississippi well. p—while mud circulation, p_s—at static temperature profile (T_f).

H, m	T, °C	T_f, °C	WBM, No. 6			OBM, No. 12		
			p, bar	p_s, bar	p-p^*, bar	P, bar	p_s, bar	p-p^*, bar
500	64.0	34.0	105.0	106.0	1.2	104.1	105.9	2.1
1,000	72.4	46.8	210.3	212.3	2.1	208.6	211.8	3.9
1,500	80.9	59.7	315.4	318.6	3.3	313.3	317.9	5.3
2,000	89.3	72.6	421.1	424.8	3.8	418.5	424.0	6.4
2,500	97.7	85.4	526.5	531.2	4.6	524.0	530.3	7.1
3,000	106.1	98.3	632.2	637.3	5.1	629.9	636.8	7.4
3,500	114.6	111.1	737.8	743.2	5.7	736.3	743.5	7.3
4,000	123.0	124.0	843.6	849.1	6.2	843.1	850.4	6.7
4,500	131.4	136.9	949.2	954.7	6.8	950.4	957.6	5.6
5,000	139.9	149.7	1055.1	1060.1	7.1	1058.2	1065.2	3.9
5,500	148.3	162.6	1160.7	1164.9	7.7	1166.6	1173.1	1.8
6,000	156.7	175.4	1266.5	1269.6	8.2	1275.5	1281.3	0.9
6,500	165.2	188.3	1372.3	1373.9	8.5	1385.1	1390.0	4.3
7,000	173.6	201.2	1477.9	1477.7	9.2	1495.3	1499.2	8.2
7,214	177.2	206.7	1523.0	1522.0	9.6	1542.6	1546.0	10.1

The parameters in Eq. (A-1) are Fluid No. 6 (WBM): $\rho_0 = 2166.3$ kg/m³, $\alpha = 4.3941 \cdot 10^{-5}$ 1/bar, $-\beta = 2.4385 \cdot 10^{-4} 1/°C$, $-\gamma = 1.3428 \cdot 10^{-6}(1/°C)^2$

Fluid No. 12 (OBM): $\rho_0 = 2166.2$ kg/m³, $\alpha = 6.5953 \cdot 10^{-5}$ 1/bar, $-\beta = 5.3037 \cdot 10^{-4}$ 1/°C, $\gamma = 1.0365 \cdot 10^{-7}(1/°C)^2$

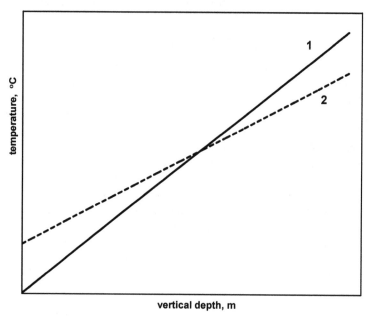

Figure A-2. Geothermal temperature (1) and drilling mud circulating temperature (2); Mississippi well, $H = 7,214$ m.

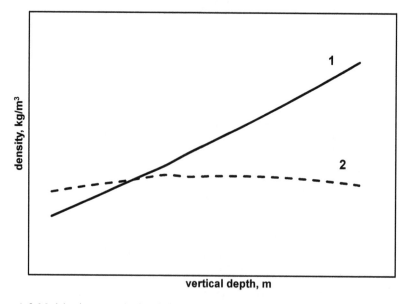

Figure A-3. Mud density versus depth. Mississippi well (1) OBM, Fluid No. 12; (2) WBM, Fluid No. 6.

A6 Hydrostatic Pressure for Water Based Fluids and Brines

For water-based muds and brines the coefficient γ in Eq. (A-1) is always a negative quantity and the integration of Eq. (A-10) presents no special problems (Babu 1996; Kårstad and Aadnøy 1998). Taking into account Eqs. (A-1) and (A-11)–(A-13) the integration of Eq. (A-10) yields (Babu 1996; Kårstad and Aadnøy 1998):

$$p = \frac{1}{\alpha} \ln \frac{1}{1 - F(h)}, \tag{A-21}$$

$$\left\{ \begin{array}{l} F(h) = B_c \alpha \rho_0 \sqrt{\pi b} \, \exp\!\left(b a^2 + b_1\right) \cdot \\ \left[\Phi\!\left(a\sqrt{b} + \dfrac{h}{2\sqrt{b}}\right) - \Phi\!\left(a\sqrt{b}\right) \right], \quad B_c = 0.052 \dfrac{\text{psi}}{\text{ppg} \cdot \text{ft}} \end{array} \right\}, \tag{A-22}$$

$$a = -b_2, \quad b = -\frac{1}{4b_3}, \tag{A-23}$$

$$b_1 = \beta\!\left(a_0 - T_s\right) + \gamma\!\left(a_0 - T_s\right)^2, \tag{A-24}$$

$$b_2 = \beta a_1 + 2a_1 \gamma (a_0 - T_s), \tag{A-25}$$

$$b_3 = \gamma a_1^2, \tag{A-26}$$

where Φ is the error function.

A7 Example of Calculation

After completion of drilling operations a 2000-m water producing well was shut-in for a long period of time. The wellbore temperatures were practically identical to undisturbed temperatures of surrounding formations. The surface temperature of formations is 10°C and the geothermal gradient is 0.045°C/m. The aquifer is located in the vicinity of the bottom-hole. The production of hot water (100°C) started at a low flow rate and the flow rate was gradually increased. Due to heat losses to the surrounding wellbore formations the outlet temperature varied between 50 and 95°C. What was the downhole hydrostatic pressure profile prior to water production? It is also required to estimate the bottom-hole pressure variation during the production period. The results of calculations after Eq. (A-21) are presented in Table A-8 and Table A-9.

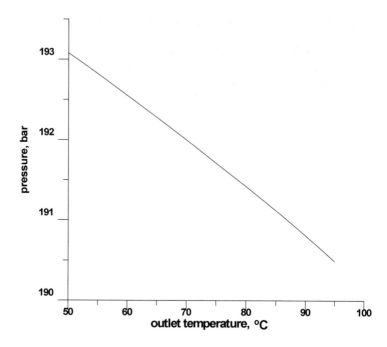

Figure A-4. Bottom-hole pressure vs. outlet water temperature.

Table A-9. The effect of water density-depth variation on hydrostatic pressure predictions.

H, m	p, bar	p-p^*, bar	ρ, kg/m^3	T, °C
200	20.66	0.009	1001	19
400	40.27	−0.020	999	28
600	59.83	−0.095	996	37
800	79.34	−0.222	993	46
1000	98.79	−0.409	990	55
1200	118.17	−0.663	987	64
1400	137.48	−0.992	983	73
1600	156.71	−1.403	978	82
1800	175.84	−1.903	974	91
2000	194.88	−2.499	968	100

References to Part II

API RP 10B, 1997. Recommended Practice for Testing Well Cements. Standard by American Petroleum Institute, USA.

API Specifications for Materials and Testing for Well Cements. 1982. API Spec. 10, First Ed., API, Dallas, USA.

Babu, D.R. 1996. Effect of p- ρ- T behavior of muds on static pressures during deep-well drilling. SPEDC, SPE Paper 27419, pp. 91–97. http://www.spe.org/publications/journals.php.

Balobayev, V.T. 1991 (in Russian). Geothermics of the Frozen Zone of the Lithosphere of the Northern Asia. Nauka, Novosibirsk.

Beck, A.E. and Balling, N. 1988. Determination of virgin rock temperatures. pp. 59–85. In: R. Haenel, L. Rybach and L. Stegena (eds.). Handbook of Terrestrial Heat-Flow Density Determination. Kluwer Academic Publ., Dordrecht/Boston/London.

Beltrami, H., Jessop, A.M. and Mareschal, J.C. 1992. Ground temperature histories in eastern and central Canada from geothermal measurements: Evidence of climate change. Palaeogeography, Palaeoclimatology and Palaeoecology (Global Planet. Change Sect.) 98, Elsevier, pp. 167–183. http://www.journals.elsevier.com.

Berthold, S. and Börner, F. 2008. Detection of free vertical convection and double-diffusion in groundwater monitoring wells with geophysical borehole measurements. Environmental Geology, 54, Springer, New York, pp. 1547–1566.

Bejan, A. 2004. Convection Heat Transfer, 3rd ed., J. Wiley and Sons, Inc., Hoboken, New Jersey.

Bogomolov, G.V., Lubimova, E.A., Tcibulya, L.A., Kutasov, I.M. and Atroshenko, P.P. 1970 (in Russian). Heat Flow of the Pripyat Through. Reports of Academy of Sciences of the Belorussia, Physical-Technical Series, Minsk 2: 97–103.

Boreholes locations and permafrost depths, Alaska, USA, from U.S. Geological Survey, 1998; http://nsidc.org/data/docs/fgdc/ggd223_boreholes_alaska.

Bradford, B.B. 1985. Attention to Primary Cementing Practices Leads to Better Jobs. Oil and Gas Jour., 21 October, pp. 59–63.

Burnham, C.W., Hollaway, J.H. and Davis, N.F. 1969. Thermodynamic properties of water to 1,000°C and 10,000 Bars. Geol. Soc. Amer. Special Paper, 32, Geological Society of America, Boulder, pp. 1–96.

Calvert, D.G. and Griffin, T.J., Jr. 1998. Determination of temperatures for cementing in wells drilled in deep water. Proceedings, SPE paper 39315 presented at the 1998 IADC/SPE Drilling Conf., 3–6 March 1998, Dallas, Texas.

Carcione, J.M. and Poletto, F. 2000. Sound velocity of drilling mud saturated with reservoir gas. Geophysics, 65, No. 2, Society of Exploration Geophysicists, Tulsa, pp. 646–651.

Carslaw, H.S. and Jaeger, J.C. 1959. Conduction of Heat in Solids. 2nd ed. Oxford University Press, London.

Cermak, V. 1971. Underground temperature and inferred climatic temperature of the past millennium. Palaeogeography, Palaeoclimatology and Palaeoecology, 10, Elsevier, pp. 1–19. http://www.journals.elsevier.com.

Cermak, V., Safanda, J. and Bodri, L. 2007. Precise temperature monitoring in boreholes: Evidence for oscillatory convection? Part I. Experiments and field data. International Journal of Earth Sciences, Springer, New York, doi: 10-1007/s00531-007-0237-4.

Cermak, V., Safanda, J. and Kresl, M. 2008. Intra-hole fluid convection: High-resolution temperature time monitoring. Jour. of Hydrology, 348, No. 3-4, Elsevier, pp. 464–479. http://www.journals.elsevier.com.

Chiu, K. and Thakur, S.C. 1991. Modeling of Wellbore Heat Losses in Directional Wells Under Changing Injection Conditions. SPE 2287, Presented at the 66th Annual Technical Conf. and Exhib. of the Society of Petroleum Engin., Dallas, TX, October 6–9.

Clauser, C. and Mareschal, J.-C. 1995. Ground Temperature History in Central Europe from Borehole Temperature Data. Geophysical Jour. International 121: 805–817. http://onlinelibrary.wiley.com/journal/10.1111/%28ISSN%291365-246X/issues.

Clow, G.D. and Lachenbruch, A.H. 1998. Boreholes locations and permafrost depths, Alaska, USA, from the U.S. Geol. Survey. http://nsidc.org/data/docs/fgdc/ggd223_boreholes_alaska/.

Covan, M. and Sabins, F. 1995. New correlations improve temperature predictions for cementing and squeezing. Oil and Gas Jour., Aug. 21, Oil and Gas Journal, Huston, pp. 53.

Dillenbeck, R.L., Heinold, T., Rogers, M.J. and Mombourquette, I.G. 2002. The Effect of cement heat hydration on the maximum annular temperature of oil and gas wells. SPE 77756 presented at the SPE Annual Technical Conf. 29 September–2 October 2002, San Antonio, Texas.

Diment, W.H. 1967. Thermal regime of large diameter borehole: Instability of the water column and comparison of air- and water-filled conditions. Geophysics, 32, Society of Exploration Geophysicists, Tulsa, pp. 720–726.

Diment, W.H. and Urban, Th. C. 1983. A simple method for detecting anomalous fluid motions in boreholes from continuous temperature logs. GRC Trans 7, Gordon Research Conferences, pp. 485–490. http://www.grc.org/home.aspx.

Djamalova, A.S. 1969 (in Russian). Heat Flow of Dagestan. Nauka, Moscow.

Dowdle, W.L. and Cobb, W.M. 1975. Static formation temperatures from well logs—an empirical method. Jour. of Petroleum Technology, Vol. 27(11), SPE, pp. 1326–1330. http://www.spe.org/publications/journals.php.

Drury, M.L. 1984. On a possible source of error in extracting equilibrium formation temperatures from boreholes BHT data. Geothermics, Vol. 13, Elsevier, pp. 175–180. http://www.journals.elsevier.com/geothermics/.

Earlougher, R.C., Jr. 1977. Advances in Well Test Analysis. SPE, New York, Dallas, Vol. 8.

Edwardson, M.L., Girner, H.M., Parkinson, H.R., Williams, C.D. and Matthews, C.S. 1962. Calculation of formation temperature disturbances caused by mud circulation. Jour. of Petroleum Technology, 14, No. 4, SPE, pp. 416–426. http://www.spe.org/publications/journals.php.

Eppelbaum, L.V. and Kutasov, I.M. 2006a. Temperature and pressure drawdown well testing: Similarities and differences. Jour. of Geophysics and Engineering, 3, No. 1, IOP Publishing, Bristol, pp. 12–20.

Eppelbaum, L.V. and Kutasov, I.M. 2006b. Determination of formation temperatures from temperature logs in deep boreholes: comparison of three methods. Jour. of Geophysics and Engineering, 3, No. 4, IOP Publishing, Bristol, pp. 348–355.

Eppelbaum, L.V. and Kutasov, I.M. 2011. Estimation of the effect of thermal convection and casing on temperature regime of boreholes—a review. Jour. of Geophysics and Engineering, 8, IOP Publishing, Bristol, R1–R10.

Eppelbaum, L.V. and Kutasov, I.M. 2013. Cylindrical probe with a variable heat flow rate: A new method for determination of the formation thermal conductivity. Central European Jour. of Geosciences, 5, No. 4, Springer, New York, pp. 570–575.

Eppelbaum, L.V., Kutasov, I.M. and Barak, G. 2006. Ground surface temperature histories inferred from 15 boreholes temperature profiles: Comparison of two approaches. Earth Sciences Research Journal, 10, No. 1, Universidad Nacional de Colombia, Bogota, pp. 25–34.

Eppelbaum, L.V., Kutasov, I.M. and Pilchin, A.N. 2014. Applied Geothermics. Springer, New York.

Eppelbaum, L.V., Modelevsky, M.M. and Pilchin, A.N. 1996. Thermal investigation in petroleum geology: the experience of implication in the Dead Sea Rift zone, Israel. Jour. of Petroleum Geology, 19, No. 4, pp. 425–444. http://onlinelibrary.wiley.com/journal/10.1111/%28ISSN%291747-5457.

Farris, F.R. 1941. A practical evaluation of cements for oil wells. Oil and Gas Co, Tulsa, Okla.

Fertl, W.H. and Wichmann, P.A. 1977. How to determine static BHT from well log data. World Oil, 184, No. 1, Gulf Publishing Company, Houston, pp. 105–106.

Gates, J.A. and Wood, R.H. 1985. Densities of aqueous solutions of NaCl, $MgCl_2$, KCl, NaBr, LiCl, and $CaCl_2$ from 0.05 to 5.0 mol kg^{-1} and 0.1013 to 40MPa at 298.15 K. Jour. of Chemical and Engineering Data, 30, No. 1, American Chemical Society, pp. 44–49. http://pubs.acs.org/journal/jceaax.

Gates, J.A. and Wood, R.H. 1989. Density and apparent molar volume of aqueous $CaCl_2$ at 323–600 K. Jour. of Chemical and Engineering Data, 34, No. 1, American Chemical Society, pp. 53–56. http://pubs.acs.org/journal/jceaax.

Gradshtein, I.S. and Ryzhik, I.M. 1965. Table of Integrals, Series and Products. Oxford University Press, London.

Gretener, P.E. 1967. On the thermal instability of large diameter wells—an observational report. Geophysics, XXXII, Society of Exploration Geophysicists, Tulsa, pp. 727–738.

Goodman, M.A., Mitchell, R.F., Wedelich, H., Galate, J.W. and Presson, D.M. 1988. Improved Circulating Temperature Correlations for Cementing, SPE paper 18029 presented at the 63rd SPE Annual Techn. Conf. and Exhib., Houston, Texas, October 2–5, 1988.

Grossman, S.I. 1977. Calculus. Academic Press, NY, San Francisco, London.

Guillot, F., Boisnault, J.M. and Hujeux, J.C. 1993. A Cementing Temperature Simulator to Improve Field Practice. Proceedings, SPE paper 25696 presented at the 1993 SPE/IADC Drilling Conf., 23–25 February, Amsterdam.

Guyod, H. 1946. Temperature Well Logging. Oil Weekly, 123, 21 Oct–16 Dec, 7 parts, Gulf Publishing Company, Houston, pp. 42.

Günzel, U. and Wilhelm, H. 2000. Estimation of the in-situ thermal resistance of a borehole using the Distributed Temperature Sensing (DTS) technique and Temperature Recovery Method (TRM). Geothermics, 29, Elsevier, pp. 689–700. http://www.journals.elsevier.com/geothermics/.

Halliburton Cementing Tables, 1979. Halliburton Company, Duncan, OK.

Hales, A.L. 1937. Convection currents in geysers. Mon. Not. Roy, Astron. Soc., Geophys. Suppl. 4: 122–131. http://mnras.oxfordjournals.org/.

Harris, M.N. and Chapman, D.S. 1995. Climate change on the Colorado Plateau of eastern Utah inferred from borehole temperatures. Jour. of Geophysical Research 100: 6367–6381. http://agupubs.onlinelibrary. wiley.com/agu/jgr/journal/10.1002/%28ISSN%292156-2202/.

Hawkins, M.F. (Jr.). 1956. A note on the skin effect. Trans. AIME, 207, American Institute of Mining, Metallurgical, and Petroleum Engineers, New York, pp. 356–357.

Hoberock, L.L., Thomas, D.C. and Nickens, H.V. 1982. Here's how compressibility and temperature affect bottom-hole mud pressure. Oil and Gas Jour., March 22, Oil and Gas Journal, Huston, pp. 159–164.

Honore, R.S., Jr., Tarr, B.A., Howard, J.A. and Lang, N.K. 1993. Cementing temperature predictions based on both downhole measurements and computer predictions: a case history. Proceedings, SPE paper 25436 presented at the Production Operations Symp., 21–23 March, Oklahoma City, OK, USA.

Huang, S., Pollack, H.N. and Shen, P.Y. 2000. Temperature trends over past five centuries reconstructed from borehole temperatures. Nature, 403, Feb. 17, pp. 756–758. http://www.nature.com/nature/journal/ v403/n6771/full/403756a0.html.

Hubbard, J.T. 1984. How temperature and pressure affect clear brines. Petrol. Eng. Intern. 56(5): 58–64.

Jacob, C.E. and Lohman, S.W. 1952. Non-steady flow to a well of constant drawdown in an extensive aquifer. Trans. Amer. Geophys. Union, August, American Geophysical Union, John Wiley and Sons, N.J., pp. 559–569.

Jaeger, J.C. 1956. Numerical values for the temperature in radial heat flow. Jour. of Math. Physics 34: 316–321. http://scitation.aip.org/content/aip/journal/jmp.

Jaeger, J.C. 1961. The effect of the drilling fluid on temperature measured in boreholes. Jour. of Geophys. Research, 66, American Geophysical Union, John Wiley and Sons, N.J., pp. 563–569.

Jones, R.R. 1986. A novel economical approach for accurate real time measurement of wellbore temperature. SPE Paper 15577, Richardson, Texas, USA.

Jorden, J.R. and Campbell, F.L. 1984. Well Logging I—Rock Properties, Borehole Environment, Mud and Temperature Logging. SPE of AIME, N.Y., Dallas, pp. 131–146.

Judge, A.S., Taylor, A.E. and Burgess, M. 1979. Canadian Geothermal Data Collection-Northern Wells 1977–78. Geothermal Series, 11, Earth Physics Branch, Energy, Mines and Resources, Ottawa.

Judge, A.S., Taylor, A.E., Burgess, M. and Allen, V.S. 1981. Canadian Geothermal Data Collection-Northern Wells 1978– 80. Geothermal Series, 12, Earth Physics Branch, Energy, Mines and Resources, Ottawa.

Kappelmeyer, O. and Haenel, R. 1974. Geothermics with Special Reference to Application. Gebruder Borntrager, Berlin.

Kårstad, E. and Aadnøy, B.S. 1998. Density behavior of drilling fluids during high pressure high temperature drilling operations. SPE Paper 47806 presented at the 1998 IADC/SPE Asia Pacific Drilling Conf., Jakarta, Indonesia, Sept. 7–9.

Korn, G.A. and Korn, T.M. 1968. Mathematical Handbook for Scientists and Engineers. 2nd Edition, McGraw-Hill Book Company, N.Y.

Kraskovskiy, S.A. 1934. Geothermal Measurements in Boreholes. Geothermal Measurements in Moscow, Transactions of the TsNIGRI, No. 8, Moscow.

Kritikos, W.P. and Kutasov, I.M. 1988. Two-Point Method for Determination of Undisturbed Reservoir Temperature. Formation Evaluation, 3, No. 1, SPE, pp. 222–226. http://www.spe.org/publications/journals.php.

Kuliev, S.M., Esman, B.I. and Gabuzov, G.G. 1968. Temperature Regime of the Drilling Wells. Nedra,Moscow.

Kutasov, I.M. 1968. Determination of time required for the attainment of temperature equilibrium and geothermal gradient in deep boreholes. Freiberger Forshungshefte, C238, pp. 55–61. http://www.geo.tufreiberg.de/psf/english/preface.htm.

Kutasov, I.M. 1976 (in Russian). Thermal Parameters of Wells Drilled in Permafrost Regions. Nedra, Moscow.

Kutasov, I.M. 1987. Dimensionless temperature, cumulative heat flow and heat flow rate for a well with a constant bore-face temperature. Geothermics, 16, Elsevier, pp. 467–472. http://www.journals.elsevier.com/geothermics/.

Kutasov, I.M. 1988. Empirical correlations determines downhole mud density. Oil and Gas Jour., December 22, Oil and Gas Journal, Huston, pp. 61–63.

Kutasov, I.M. 1988. Prediction of permafrost thickness by the "Two Point" Method. Proceed. of the 5th Intern. Conf. on Permafrost, Vol. 2, Tapir Publ., Trondheim, Norway, pp. 965–970.

Kutasov, I.M. 1989a. Application of the Horner method for a well produced at a constant bottomhole pressure. Formation Evaluation, March, SPE, pp. 90–92. http://www.spe.org/publications/journals.php.

Kutasov, I.M. 1989b. Water FV factors at higher pressures and temperatures. Oil and Gas Jour., March 20, Oil and Gas Journal, Huston, pp. 102–104.

Kutasov, I.M. 1991. Correlation simplifies obtaining brine density. Oil and Gas Jour., August 5, Oil and Gas Journal, Huston, pp. 48–49.

Kutasov, I.M. 1998. Program analyses step-pressure data. Oil and Gas Jour., Jan. 5, Oil and Gas Journal, Huston, pp. 43–46.

Kutasov, I.M. 1999. Applied Geothermics for Petroleum Engineers. Elsevier, Amsterdam – NY – London,Ser: Development in Petroleum Science, pp. 48.

Kutasov, I.M. 2002. Method corrects API bottom-hole circulating-temperature correlation. Oil and Gas Jour., July 15, Oil and Gas Journal, Huston, pp. 47–50.

Kutasov, I.M. 2003. Dimensionless temperature at the wall of an infinite long cylindrical source with a constant heat flow rate. Geothermics, 32, Elsevier, pp. 63–68. http://www.journals.elsevier.com/geothermics/.

Kutasov, I.M. 2007. Determination of calcium chloride brine concentration required to provide pressure overbalance. Jour. of Petroleum Science and Engineering, 58, Elsevier, pp. 133–137. http://www.journals.elsevier.com/geothermics/.

Kutasov, I.M. 2007. Dimensionless temperatures at the wall of an infinitely long, variable-rate, cylindrical heat source. Geothermics, 36, Elsevier, pp. 223–229. http://www.journals.elsevier.com/geothermics/.

Kutasov, I.M. 2015. Analyzing the pressure response during the afterflow period—Determination of the formation permeability and skin factor. Proceed. of the World Geothermal Congress, Melbourne, Australia, pp. 1–12.

Kutasov, I.M., Atroshenko, P.P. and Tcibulya, L.A. 1971. The use of temperature logs for predicting the static formation temperature. Reports of Academy of Sciences of Belorussia, Minsk 15(3): 266–269.

Kutasov, I.M. and Devyatkin, V.N. 1964. Experimental investigation of temperature regime of shallow convective holes. pp. 143–150. *In*: Thermal Processes in Frozen Rocks, Nauka, Moscow.

Kutasov, I.M. and Devyatkin, V.N. 1973. Effect of Free Heat Convection and Casing on the Temperature Field in Boreholes. pp. 99–106. *In*: Heat Flows from the Crust and Upper Mantle of the Earth, Nauka, Moscow.

Kutasov, I.M. and Devyatkin, V.N. 1977. Experimental investigation of temperature regime of shallow convective holes. CRREL Draft Translation, 589, March, 1977. http://century65clinic.thecoolcatalog.com/.

Kutasov, I.M. and Eppelbaum, L.V. 2003. Prediction of formation temperatures in permafrost regions from temperature logs in deep wells—field cases. Permafrost and Periglacial Processes, 14, No. 3, John Wiley and Sons, N.J., pp. 247–258.

Kutasov, I.M. and Eppelbaum, L.V. 2005a. Drawdown test for a stimulated well produced at a constant bottomhole pressure. First Break 23(2): 25–28. http://onlinelibrary.wiley.com/journal/10.1111/%28 ISSN%291365-2397/issues.

Kutasov, I.M. and Eppelbaum, L.V. 2005b. Determination of formation temperature from bottomhole temperature logs—a generalized Horner method. Jour. of Geophysics and Engineering, 2, IOP Publishing, Bristol, pp. 90–96.

Kutasov, I.M. and Eppelbaum, L.V. 2007. Temperature well testing—utilization of the Slider's method. Jour. of Geophysics and Engineering, 4, No. 1, IOP Publishing, Bristol, pp. 1–6.

Kutasov, I.M. and Eppelbaum, L.V. 2008. Designing an interference well test in a geothermal reservoir. Proceed. of 32rd Workshop on Geothermal Reservoir Engineering, Stanford Univ., SGP-TR-185, Stanford, California, Jan. 28–30.

Kutasov, I.M. and Eppelbaum, L.V. 2009. Estimation of the geothermal gradients from single temperature log-field cases. Jour. of Geophysics and Engineering, 6, No. 2, IOP Publishing, Bristol, pp. 131–135.

Kutasov, I.M. and Eppelbaum, L.V. 2010. A new method for determination of formation temperature from bottom-hole temperature logs. Jour. of Petroleum and Gas Engineering 1(1): 1–8. http://www.academicjournals.org/journal/JPGE.

Kutasov, I.M. and Eppelbaum, L.V. 2011. Recovery of and superdeep wells: Utilization of measurements while drilling data. Proceed. of the 2011 Stanford Geothermal Workshop, Stanford, USA, SGP-TR-191, pp. 1–7.

Kutasov, I.M. and Eppelbaum, L.V. 2012a. New method evaluated efficiency of wellbore stimulation. Oil and Gas Journal, 110, No. 8, Oil and Gas Journal, Houston, pp. 22–24.

Kutasov, I.M. and Eppelbaum, L.V. 2012b. Geothermal investigations in permafrost regions—the duration of temperature monitoring after wellbores shut-in. Geomaterials, 2, No. 4, Scientific Research Publishing, pp. 82–93. http://www.scirp.org/journal/gm/.

Kutasov, I.M. and Eppelbaum, L.V. 2012c. Cementing of geothermal wells—radius of thermal influence. Proceed. of the 2012 Workshop on Geothermal Reservoir Engineering, Stanford Univ., Stanford, California, January 30–February 1.

Kutasov, I.M. and Eppelbaum, L.V. 2013a. Cementing of casing—temperature increase at cement hydration. Proceed. of the 2013 Stanford Geothermal Workshop, Stanford, USA, pp. 1–6.

Kutasov, I.M. and Eppelbaum, L.V. 2013b. Cementing of casing: the optimal time lapse to conduct a temperature log. Oil Gas European Magazine, 39, No. 4, OilGasPublisher.de, pp. 190–193. http://www.oilgaspublisher.de/.

Kutasov, I.M. and Eppelbaum, L.V. 2014. Temperature regime of boreholes: Cementing of production liners. Proceed. of the 2014 Stanford Geothermal Workshop, Stanford, USA, pp. 1–5.

Kutasov, I.M. and Eppelbaum, L.V. 2015. Wellbore and Formation Temperatures during Drilling, Shut-in and Cementing of Casing. Proceed. of the World Geothermal Congress, Melbourne, Australia, pp. 1–12.

Kutasov, I.M., Caruthers, R.M., Targhi, A.K. and Chaaban, H.M. 1988. Prediction of downhole circulating and shut-in temperatures. Geothermics, 17, Elsevier, pp. 607–618. http://www.journals.elsevier.com/geothermics/.

Kutasov, I.M. and Kagan, M. 2000. Forecast of injectivity for a well with a constant bottomhole pressure. Proc. 22st NZ Geothermal Workshop, University of Auckland, pp. 233–235.

Kutasov, I.M. and Kagan, M. 2001. Determination of the skin factor for a well produced at a constant bottomhole pressure. SPE Paper 67241, Proceed. of the SPE Production and Operations Symposium, Oklahoma City, Oklahoma, pp. 24–27.

Kutasov, I.M. and Kagan, M. 2003a. Determination of the skin factor for a well produced at a constant bottomhole pressure. Jour. of Energy Resources Techn., 125, American Society of Mechanical Engineers, pp. 61–63. http://energyresources.asmedigitalcollection.asme.org/journal.aspx.

Kutasov, I.M. and Kagan, M. 2003b. Cylindrical probe with a constant temperature—determination of the formation thermal conductivity and contact thermal resistance. Geothermics, 32, Elsevier, pp. 187–193. http://www.journals.elsevier.com/geothermics/.

Kutasov, I.M., Lubimova, E.A. and Firsov, F.V. 1966 (in Russian). Rate of recovery of the temperature field in wells on Kola Peninsula. pp.74–87. *In*: Problems of the Heat Flux at Depth. Nauka, Moscow.

Kutasov, I.M. and Seman, T. 2001. Determination of brine density for HPHT wells. Oil and Gas Jour., Dec. 24, Oil and Gas Journal, Huston, pp. 55–57.

Kutasov, I.M. and Strickland, D.G. 1988. Allowable shut-in is estimated for wells in permafrost. Oil and Gas Journal, September 5, Oil and Gas Journal, Huston, pp. 55–60.

Kutasov, I.M. and Targhi, A.K. 1987. Better deep-hole BHCT estimations possible. Oil and Gas Journal, 25 May, Oil and Gas Journal, Huston, pp. 71–73.

Lachenbruch, A.H. and Brewer, M.C. 1959. Dissipation of the temperature effect of drilling a well in Arctic Alaska. U.S. Geological Survey Bull. 1083-C, U.S. Geological Survey, pp. 74–109. https://pubs.er.usgs.gov/browse/usgs-publications/B.

Lachenbruch, A.H., Cladouhos, T.T. and Saltus, R.W. 1988. Permafrost temperature and the changing climate. Proceed. of the Fifth International Conf. on Permafrost, Vol. 3, Tapir Publishers, Trondheim, Norway, pp. 9–17.

Lee, J. 1982. Well Testing. Texas, SPE Monograph Series. http://www.spe.org/publications/journals.php.

Majorowicz, J.A., Jones, F.W. and Judge, A.S. 1990. Deep subpermafrost thermal regime in the Mackenzie Delta Basin, northern Canada—Analysis from petroleum bottom-hole temperature data. Geophysics, 55, Society of Exploration Geophysicists, Tulsa, pp. 362–371.

McMordie, W.C., Jr., Bland, R.G. and Hauser, J.M. 1982. Effect of temperature and pressure on the density of drilling fluids. SPE Paper 11114. Trans. of the SPE 57th Ann. Fall Technical Conf. and Exhib., New Orleans, LA, 26–29 Sept., 1982.

Melnikov, P.I., Balobayev, V.T., Kutasov, I.M. and Devyatkin, V.N. 1973. Geothermal studies in Central Yakutia. Intern. Geol. Rev. 16(5): 565–568. http://www.tandfonline.com/toc/tigr20/current.

Mufti, I.R. 1971. Geothermal aspects of radioactive waste disposal into the subsurface. Jour. of Geophys. Research, 76, American Geophysical Union, John Wiley and Sons, N.J., pp. 8563–8568.

New Cement Test Schedules Issued. 1977. Oil & Gas Jour., July 25, pp. 179–182.

Osterkamp, T.E. 1984. Response of Alaska permafrost to climate. Proceed. of Fourth Intern. Permafrost Conf., 17–22 July, 1983, Fairbanks, Alaska, Nat. Acad. of Sci., Washington, D.C., pp. 145–151.

Ostroumov, G.A. 1952 (in Russian). Free Convection Under the Conditions of the Internal Problem. Gostekhizdat, Moscow.

Pfister, M. and Rybach, L. 1995. High-resolution digital temperature logging in areas with significant convective heat transfer. Geothermics, 24, No. 1, Elsevier, pp. 99–100. http://www.journals.elsevier.com/geothermics/.

Potter, R.W. II and Brown, D.L. 1977. The volumetric properties of aqueous sodium chloride solutions from 0° to 500 +C at pressures up to 2000 bars based on a regression of available data in literature. In: Pleminary Tables for NaCl Solutions, Geol. Survey Bull., 1421-C, U.S. Gov. Printing Office, Washington, D.C.

Powell, W.G., Chapman, D.S., Balling, N. and Beck A.E. 1988. Continental Heat-Flow Density. pp. 167–222. In: Haenel, R., Rybach, L. and Stegena, L. (eds.) Handbook of Terrestrial Heat-Flow Density Determination. Kluwer Acad. Publishers, Dordrecht/Boston/London.

Proselkov, Yu. M. 1975 (in Russian). Heat Transfer in Wells. Nedra, Moscow.

Ramey, H.J. (Jr.). 1962. Wellbore heat transmission. Jour. of Petroleum Technology, 14, No. 4, SPE, pp. 427–435. http://www.spe.org/publications/journals.php.

Raymond, L.R. 1969. Temperature distribution in a circulating drilling fluid. Jour. of Petroleum Technology, SPE, 21, No. 3, pp. 333–341. http://www.spe.org/publications/journals.php.

Romero, J. and Loizzo, M. 2000. The importance of hydration heat on cement strength development for deep water wells. SPE paper 62894 presented at the 2000 SPE Annual Technical Conf. and Exhib., Dallas, Texas, 1–4 October.

Roux, B., Sanyal, S.K. and Brown, S.L. 1980. An improved approach to estimating true reservoir temperature from transient temperature data. SPE paper 8888 SPE California Regional Meeting (Los Angeles, 9–11 April 1980).

Sammel, E.A. 1968. Convective flow and its effect on temperature logging in small-diameter wells. Geophysics, 33, Society of Exploration Geophysicists, Tulsa, pp. 1004–1012.

Santoyo, E., Garcia, A., Espinosa, G., Gonzalez-Partida, E. and Viggiano, J.C. 2000. Thermal evaluation study of the LV-3 well in the Tres Virgenes geothermal field. Mexico Proceedings World Geothermal Congress (Kyushu-Tohoku, Japan, May 28–June 10, 2000), pp. 2177–2182.

Sass, J.H., Kennelly, J.P. (Jr.), Wendt, W.E., Moses, T.H. (Jr.) and Ziagos, J.P. 1981. In-situ determination of heat flow in unconsolidated sediments. Geophysics, 46, No. Society of Exploration Geophysicists, Tulsa, pp. 176–83.

Schoeppel, R.J. and Gilarranz, S. 1966. Use of well log temperatures to evaluate regional geothermal gradients. Jour. of Petroleum Technology SPE 18(6): 667–673. http://www.spe.org/publications/journals.php.

Sengul, M.M. 1983. Analysis of step-pressure tests. SPE paper 12175, 58th Annual technical Conf. and Exhibition (San Francisco, October 5–8, 1983).

Shell, F. and Tragesser, A.F. 1972. API is Seeking More Accurate Bottom Hole Temperatures. Oil Gas J., 10 July 10: 72–79.

Shen, P.Y. and Beck, A.E. 1992. Paleoclimate change and heat flow density inferred from temperature data in the Superior Province of the Canadian Shield. Palaeogeography, Palaeoclimatology and Palaeoecology (Global Planet. Change Sect.), 98, Elsevier, pp. 143–165. http://www.journals.elsevier.com.

Somerton, W.H. 1992. Thermal Properties and Temperature Related Behaviour of Rock/Fluid Systems, Developments in Petroleum Science. Elsevier. http://www.journals.elsevier.com.

Sorelle, R.R., Jardiolin, R.A. and Buckley, P. 1982. Mathematical field model predicts downhole density changes in static drilling fluids.: SPE Paper 11118, Presented at the 57th SPE Annual Fall Technical Conf. and Exhib., New Orleans, LA, 26–29 September.

Stephens, M. and Lau, H.C. 1998. Completion fluids. pp. 325–343. *In*: M.J. Economides, L.T. Watters and S. Dunn-Norma (eds.). Petroleum Well Construction, John Wiley & Sons, N.J.

Sump, G.D. and Williams, B.B. 1973. Prediction of wellbore temperatures during mud circulation and cementing operations. Jour. of Engineering for Industry, 95, Ser B, No. 4, American Society of Mechanical Engineers, New York, pp. 1083–1092.

Taylor, A.E. and Judge, A.S. 1976. Canadian Geothermal Data Collection—Northern Wells 1975, Geothermal Series 6, Earth Physics Branch, Energy, Mines and Resources, Ottawa.

Taylor, A.E. and Judge, A.S. 1977. Canadian Geothermal Data Collection-Northern Wells 1976–77, Geothermal Series, 10, Earth Physics Branch, Energy, Mines and Resources, Ottawa.

Taylor, A.E. 1978. Temperatures and Heat Flow in a System of Cylindrical Symmetry Including a Phase Boundary. Geothermal Series, 7, Ottawa, Canada.

Taylor, A.E., Burgess, M., Judge, A.S. and Allen, V.S. 1982. Canadian Geothermal Data Collection— Northern Wells 1981, Geothermal Series 13, Earth Physics Branch, Energy, Mines and Resources, Ottawa.

The Engineering Toolbox 20.02.2010. http://www.engineeringtoolbox.com/water-thermal-properties-d_162. html.

Thomas, D.C., Atkinson, G. and Atkinson, B.L. 1984. Pressure and temperature effects on brine completion fluid density. SPE Paper 12489 presented at the Formation Damage Control Symp., Bakersfield, CA, February 13–14.

Timko, D.J. and Fertl, W.H. 1972. How downhole temperatures and pressures affect drilling. World Oil, 175, Gulf Publishing Company, Houston, pp. 73–78.

Tsytovich, N.A. 1975. The Mechanics of Frozen Ground. Script a Book Co., Washington, D.C.

Van Everdingen, A.F. and Hurst, W. 1949. The application of the Laplace transformation to flow problems in reservoirs. Trans. AIME, 186, American Institute of Mining, Metallurgical, and Petroleum Engineers, New York, pp. 305–324.

Venditto, J.J. and George, C.R. 1984. Better wellbore temperature data equals better cement jobs. World Oil, February, Gulf Publishing Company, Houston, pp. 47–50.

Uraiet, A.A. and Raghavan, R. 1980. Unsteady flow to a well producing at constant pressure. Jour. of Petroleum Technology, SPE, pp. 1803–1812. http://www.spe.org/publications/journals.php.

Waples, D.W. and Ramly, M. 2001. A statistical method for correcting log-derived temperatures. Petroleum Geoscience 7(3): 231–240. https://www.eage.org/?evp=7987.

Waples, D.W., Pachco, J. and Vera, A. 2004. A method for correcting log-derived temperatures deep wells calibrated in the Gulf of Mexico. Petroleum Geoscience 10: 239–245. https://www.eage. org/?evp=7987.

Waterloo Maple. 2001. Maple 7 Learning Guide, Waterloo Maple Inc., Waterloo, Canada.

Wilhelm, H., Baumann, C. and Zoth, G. 1995. Some Results of Temperature Measurements in the KTB Main Borehole, Germany. Geothermics, 24, No.1, Elsevier, pp. 101–113. http://www.journals.elsevier. com/geothermics/.

Wimby, J.M. and Berntsson, T.S. 1994. Viscosity and density of aqueous solutions of LiBr, LiCl, $ZnBr_2$, $CaCl_2$ and $LiNO_3$. 1. Single Salt Solutions. Jour. of Chemical and Engineering Data, 39, No. 1, American Chemical Society, pp. 68–72. http://pubs.acs.org/journal/jceaax.

Wisian, K.W., Blackwell, D.D., Bellani, S., Henfling, J.A., Norman, R.A., Lysne, P.C., Forster, A. and Schrotter, J. 1998. Field comparison of conventional and new temperature logging systems. Geothermics, 27, No. 2, Elsevier, pp. 131–141. http://www.journals.elsevier.com/geothermics/.

Wooley, G.R., Giussani, A.P., Galate, J.W. and Wederlich, H.F. (III). 1984. Cementing temperatures for deep-well production liners. SPE paper 13046, 59th Annual Technical Conf. and Exhibition (Houston, Texas, 16–19 September 1984).

Zschocke, A. 2005. Correction of non-equilibrated temperature logs and implications for geothermal investigations. Jour. of Geophysics and Engineering, 2, IOP Publishing, Bristol, pp. 364–371.

Index